Nutritional Genomics—A Consumer's Guide to How Your Genes and Ancestry Respond to Food

Nutritional Genomics—A Consumer's Guide to How Your Genes and Ancestry Respond to Food

Tailoring What You Eat to Your DNA

Anne Hart

iUniverse, Inc.
New York Lincoln Shanghai

Nutritional Genomics—A Consumer's Guide to How Your Genes and Ancestry Respond to Food
Tailoring What You Eat to Your DNA

iUniverse, Inc.

For information address:
iUniverse, Inc.
2021 Pine Lake Road, Suite 100
Lincoln, NE 68512
www.iuniverse.com

ISBN: 0-595-29067-1

Printed in the United States of America

Contents

▼

Introduction

DNA Testing for Nutritional Genomics and Ancestry

Tailor what you eat to your DNA. This book is a consumer's guide to nutritional genomics. It's a guide to how your particular genes and ancestry respond to food—selected foods tailored or prescribed to nourish your individual genetic profile known as your *genetic expression*. Consumer involvement in nutritional genomics is important.

Nutritional genomics, often abbreviated as 'nutrigenomics' is about increasing that success rate. How will science working together with the consumer tackle the issues confronting us as the population ages? Consumer involvement can democratize the science of nutritional genomics by improving diets for better health. You can ask to work on ethics boards or create your own. How is discovering deep ancestry through DNA testing related to the ways that food affects your health?

Ancestry and diet are linked by biology, culture, and choices. It's all about collaborating with your genes. Do you choose your food by habit or biology? Consumers need a guide to DNA testing for nutritional genomics as well as for ancestry and family history. Specific genetic variants *interact* and relate to nutrition.

Learn to interpret the expression of your genes before you count your calories. If you're supposed to eat 'bright' for your 'genotype,' then you begin by mapping your genetic expression and learning how the raw data applies in a practical way to what you consume. This means genetic testing, interpretation, and application to food.

Nutritional Genomics for the Consumer

How are you managing your gene expression? In what direction are you moving? How do you make more intelligent *choices* of food to *nourish* your individual *genotype*? What is meant by *intelligent foods* that target and nourish *specific* genes?

Clinical dietitians and nutritionists, by allying with molecular geneticists, genetics counselors, physicians, molecular *genealogists*, family historians, phenomics professionals, nutritional and medical anthropologists, and archaeogeneticists are *collaborating* with consumers of genetics testing, but what are they really *sharing*?

If not so much raw materials such as DNA from donors, is shared, then how about *access* to information—databases and various discussion forums online and e-mailing lists equally open to consumers, licensed healthcare providers, and research scientists? Who controls access to new research—the consumers, the corporations, or the scientists?

Can the average consumer afford to find out what to eat for improved health and nourishment based upon tests of genetic expression? Can consumers *override* any inherited risks revealed in the genetic signature with foods and nutraceuticals individually tailored? What does it mean to eat 'smarter' foods that target specific genes compared to eating more intelligently regarding choice?

Scientists compare genetic distances between populations by comparing the frequencies of forms of genes called 'alleles.' Mutant alleles can be mapped as population genetics markers. Some, but not all mutations in genes may put you at risk for certain chronic diseases if you eat the wrong foods for your genotype. The solution is to eat more 'intelligent' foods customized to your individual genetic profile.

Research also looks at rare alleles. Their rarity gives them special power as markers of genetic similarity. There's a good chance two identical mutant alleles share a common origin. You can map genes for ancestral origin, migrations, or to reveal risks of disease depending upon which genes you map.

This book is for beginners with no science background. It's a consumer's guide book to nutritional genomics—genetic testing and profiling for foods tailored to your genotype and ancestry. The chapters also are about how to interpret DNA testing for family history and ancestry.

How do you as a consumer, not a scientist, choose the smartest food tailored to your genetic signature? How do you interpret your DNA test results for ancestry or family history? What is the *link* between tailoring your foods to your genetic expression and tracing your ancestry though DNA testing? And what genes are tested for either reason? How do you bridge the gap between nutritional genomics profiling and testing DNA for deep ancestral origins?

Does ethnicity play any role in tailoring your food and nutraceuticals, drug dosages, or healthcare? How much can the average consumer self-educate and/or start a private DNA bank for a consumer or patient group? How do you raise funds, contract with research scientists, and form or serve groups needing their DNA researched for specific reasons? How does learning how to interpret the results of your DNA tests for ancestry relate to understanding genetic tests for cardiovascular or other inheritable risks?

Start researching on your own what you need to know as a consumer to have more *choices* in customizing foods for your genetic signature—your genotype. What are some realistic applications of genetic testing and profiling?

This book will lead you to find out more about taking control of what happens to DNA that you may donate for research. You'll find out how to be in charge of your own nourishment and nutrition. Genetic profiling helps you to customize what you eat. How do you nourish your body? What can your genes reveal to you through genetic testing and profiling? It's your private information and should remain private. A good place to release it finally would be in a time capsule and history scrapbook for your heirs. Here are how some branches of human genetic history are linked to your nutrition, ancestry, and most of all nourishment.

Prosopography is all about human history and genes that travel because your genes have both a cultural and a biological component. The cultural component includes onomastics which is the study of the origin of a name and its geographical and historical utilization. Put all these branches of molecular genetics together with molecular genealogy. Add nutritional genomics, and what do you have? Knowledge of how every molecule in your body responds to certain foods, lifestyles, exercise, adverse reactions to drug dosages (pharmacogenetics), and an open door to learning how your respond to nutrition and nourishment.

Multidisciplinary nutrition research and collaboration is necessary for nutritional genomics to bring together diverse expertise. Scientists working in the disciplines of nutrition, biochemistry, and genetics need to share, collaborate, and interface in this field. If scientists are more concerned about positioning themselves first in publishing their research and won't share DNA with all scientists, how can research ever move forward?

You might want to read "The Metabolic Basis of Inherited Disorders," 6th ed. McGraw-Hill, New York: 2649-2680, 1989. Then compare the latest research in nutritional genomics on how smart foods (foods tailored to your genetic signature) influence risk of chronic disease. The longer science studies the entire genome (rather than the specific SNPs for certain chronic diseases) the more information will be forthcoming on how food and lifestyle influence your health based on the genes you inherited.

CHAPTER 1

▼

FOODS TAILORED TO YOUR GENOTYPE

How do your genes respond to what you eat? How many diet-by-DNA book titles are there? Books on smarter foods? Tailored menus? Extracts of plants? DNA tests for ancestry? Ancestry and eating? Genes are distributed, function, and work in such ways that nearly every reasonable diet could work well in about six percent of the population, according to Dr. Frederic D. Abramson, CEO, Alpha-Genics, Inc.

Are you eating smart foods—foods tailored to your genotype—DNA, your ancestry, and your entire genome of genes? Are you ready to get a picture of your response to your nutrition? How can you eat to nourish your genotype? According to Genomics 120, a science, nutrition, and health Web site at: http://g120.com/products/genomicscience.html, are you wondering why in the United States currently only 50,000 people out of some 280 million live to be even 100 years old, or that your body may be aging nearly twice the rate it should be because you're eating the wrong food for your genetic signature?

There is a strong connection between nutrition and genotype, especially in regards to your cardiovascular and central nervous system health. So you need to tailor foods intelligently to your genetic expression. The media buzz about 'intelligent' foods or 'smart' foods really means eating clean, safe, whole foods based on what your individual genes need to thrive. Not all your genes are tested. You might start your food research at the Web site of Food Resource, a source of science-based and business savvy information for the food industry at: http://food.oregonstate.edu/nutri.html.

What happens when diet books for your condition aren't working for you? Maybe salt restriction isn't working but exercise is for your condition. How do your genes respond to nutrition and *nourishment*? Are your genes intelligent, conscious, and communicating with you about their nutritional needs? If they are, so are the foods you eat. Your genes interact and collaborate as a team.

The language of communication is written in the human genome, in your individual genetic signature—in your DNA, in particular SNPs, and in all your genes and cellular material. Even your blood type is expressed in all the cells of your body. How does all this information signal you about what 'smart' foods and nutraceuticals to choose in order to help prevent or delay chronic disease for which your genes may put you at risk?

A slogan reads, "Smart foods for intelligent people." Nutritional genomics is a buzz word in the news. Testing DNA for ancestry also bridges gaps in regard to customizing smarter foods to your genotype. Phenomics is about customized healthcare and medicine tailored to your genetic profile. Prosopography is an independent science of social history embracing genealogy, onomastics and demography. If you're interested in metabolic typing, one Web site for Personal Metabolic Typing is at: http://www.wholebodyhealth.us/. Dentists may be interested to know that gum disease is genetic and may be caused by a genetic predisposition to diabetes, heart disease, or low birth weight. A genetic profile on patients with deep pockets of gum disease might be useful. Check out the Holistic Dental Network at: http://holisticdentalnetwork.com/products_services.php.

Cracking the human genome code is so new and tests so costly. Currently only certain genetic markers are tested. The genetic signatures tested include genes that tell you about risk for certain diseases. Nutritional genomics as a field of research also is abbreviated as a generic term to read 'nutrigenomics.'

Without testing all the genes, how can you know about all the diseases for which you may be at risk? And without knowing all the information that every one of your gene's reveals, how can you develop a plan to override your genetic risks by nourishing your genes with what they need to stay healthier? **Here is how some scientists answered these questions.**

According to Dr. Fredric D. Abramson, Ph.D, S.M., President and CEO of *AlphaGenics, Inc.*, "The key to using diet to manage genes and health lies in managing gene expression (which we call the Expressitype). Knowing your genotype merely tells you a starting point. Genotype is like knowing where the entrance ramps to an interstate can be found. They are important to know, but tell you absolutely nothing about what direction to travel or how the journey will go. That is why Expressitype must be the focus." You can contact AlphaGenics, Inc. at: http://www.Alpha-Genics.com or write to: Maryland Technology Incubator, 9700 Great Seneca Highway, Rockville, MD 20850.

Alpha Genics, Inc. is a nutrigenomics science company. A sidebar on the company's Web site from Dr. Fredric Abramson, CEO reads (reprinted here with permission), "We are about to see a revolution in our concept of diet. Each of us is a unique organism and for the first time in human history, genetic research is confirming that one diet is not optimum for everyone.

Science is discovering that each individual's DNA processes food and nutritious supplements in a unique way. Through the development of a cutting-edge DNA analytical system and consumer guidance, Alpha Genics will be able to tune nutrition to meet the needs of each individual resulting in optimum health, peak performance, and enhanced creativity." What I also like about Alpha-Genics Inc. is that they have an independent, separate Ethics Board. Check out that Web site at: http://www.alpha-genics.com/ethics_board.php.

It is not part of the regular Board of Directors. It has five members: three outsiders, one representative from the Board of Directors, and one representative of the employees. The Ethics Board has no veto power, but has a seat on the Board of Directors. Compensation for the Ethics Board members comes through a blind trust, which means the Ethics Board has neither control nor knowledge of how the funds develop.

"I created this because I think companies need to have independent voices to provide reality checks," says Dr. Fredric Abramson. "It is something like that scene in Patton when he talks about the Roman conqueror returning home to glory, with someone standing just beside him reminding him that fame is fleeting. An independent Ethics Board helps us make better choices."

Consumers can bridge the gap between ethics boards and the media by acting as liaisons, ombudsmen, lobbyists, trustees, recruiters, communicators, independent board members, fee-for-service contractors, industry watchers, or volunteers. Get involved in the nutritional genomics industry.

You even can put together coporate gift baskets full of nutritional genomics products or samples. Throw a nutritional genomics party in any home, office, or meeting pace, in church basements, teacher's lounges after school, or at conventions. Make tape recordings about nutritional genomics and post the radio-length broadcasts to your Web site. So many news stories in the media give the impres-

sion that the average consumer will have to *wait a decade* for genomic testing to
be applied to customized foods.

For example, the New York Times Magazine published an article by NY
Times Magazine writer Bruce Grierson titled, "What Your Genes Want You to
Eat: New Way to Look at Disease." The article also appeared in the Sacramento
Bee, a daily newspaper, on Sunday July 13, 2003 on the Science page. Sundays
are great for reading the Science page in the Forum section. There's time to read
the cutting-edge science articles, many of which are reprinted from other, major
urban publications. It's a family tradition for four decades.

I was impressed by the media buzz around the relatively new field of nutritional genomics. Last year the media buzz circled around testing DNA for
ancient ancestry and genealogy, during which I took several DNA tests for ancestry and enjoyed the results. This year DNA and diet is fast-track news. DNA and
foods also tie into food safety and security issues.

I like to know that when I go into a health food store and buy a package of
imported powdered ginger under the title of herbs or botanicals that it's clean
and contains no toxic pesticides, residues, bad bacteria, or unsafe chemicals. I
often thought about who inspects these imports and do they rush through, take
enough time, or have enough staff? If I buy fresh ginger root, I'm concerned
whether it's organic or still full of pesticide residue. On the other hand, who has
time to think these thoughts?

It all came rushing back when I went into a health food store last month,
walked past an open bin where a child about twelve years old had just put her
hand into the cous-cous, grabbed a fist full of the grain-like pasta, tossed it into
her mouth, chewed it, spit it back into her palm and replaced the cous-cous that
she had just spit into her hand, back into the bin with the rest of the grain. Her
mother was busy looking at other products.

She didn't realize I was standing beside her ready to take a scoop myself of the
oat groats in the next bin. The scoop hit the floor, and she picked it up and
replaced it. The cashier was in the front of the store, the manager in the back
room. Why couldn't the store change the bins so that nobody else could spit into
the grain? When I brought this to the cashier's attention, she shrugged, looked
down, and acted as if nothing had happened. I wondered at that instant, if this is
how one consumer's word is received, how will the public perceive us?

There's power in numbers, in grouping together. I wished I had my video
camera at that moment. There's also power in the the media—the reputable,
credible media that bridges gaps between science and the public. It's time for

some consumers to become "media people." Let's watch the watchers and look at the media.

The article titled, "What Your Genes Want You to Eat: New Way to Look at Disease," had its first opening sentence beginning with a trip to the "'diet doc' circa 2013." I eagerly wanted this kind of testing to be available now, not in 2013. So I turned to Dr. Frederic Abramson, CEO of Alpha-Genics Incorporated to interview by email on his views. Dr. Abramson also teaches part-time at Johns Hopkins University, in the graduate program in Biotechnology.

I asked Dr. Frederic Abramson, CEO of Alpha-Genics Incorporated the following questions in an e-interview. Here are his answers.

1. When do you think genomic testing might be available to most consumers?

This is an important question. From a practical perspective, genomic testing—for part of a person's total genome—is possible today. We can test for several thousand genes right now. So this leads to two subquestions. One, when will testing of an entire genome be possible? And two, when will low-cost testing be available.

Experts argue about whether we have 30,000 or 70,000 genes, or somewhere in between. Regardless of the number, we are five years away from a comprehensive full genotype test of all genes in a person. By genotype, I mean the identification of which genes a person has.

But it is not your genotype which determines things. It is the work that your specific genes do. Think of genotype as the location of exit ramps on an interstate. You need to know where these are, but they tell you nothing about where you are going and what the journey will be like.

To identify these, we must look at each gene's level of activity, called "gene expression." We call the gene expression the Expressitype. Gene expression changes over time for many genes. How do we know this? Because we age. We start as children, go though puberty, become adults, and then start declining. All of this is substantially under genetic control. Some people age faster than other. It's in their genes.

Right now testing costs a fair amount. And seldom is gene testing covered by insurance. But over time, the technology advances will enable very low cost tests. For example, measuring gene expression in several thousand genes can cost between $800 to $3,000, depending on who does it. But a Japanese company is working on a test that will end up costing less than $100 for 900 genes. Thus, one thing we can be sure of is that the costs will drop. Just like computers, VCR's and microwaves.

We are working to bring the test cost to be under $1,000, with a monthly follow-up of around $79. The monthly fee lets you contact us by phone or email anytime to ask whether something you want to eat might help you or not.

We are working with Carnegie Mellon University in Pittsburgh to develop a small implantable device that will measure vital chemicals in the blood, and send signals outside the body. This will let us track what is happening in a person around the clock, every day, with much more accuracy and less guesswork.

2. What do you think is the most important area of research in nutritional genomics today?

The most important area is to identify how the dietary system, which is composed of hundreds and thousands of chemicals in varying dosages, interacts with the thousands of genes in the genome to produce health, or illness. This is generally called "systems biology." Basically, we can no longer look at single things for easy answers. It won't be just a question of whether you eat blueberries or bananas or rice, but what balance of each of these you eat over time.

This points to the second area, to understand the dynamics of how changes in diet influence the work each person's genes is doing. The value of systems biology is that ultimately, we will be able to identify individualized responses to diet, based on genetic composition.

3. Do you have any advice for those who are looking for tailored diets for specific conditions—if genotype testing is not available today, what is the next best thing for the average consumer who has already had a DNA test of merely the mtDNA or Y chromosome for ancestry?

Ironically, many of the folk suggestions about diet weren't far wrong. So first look to your family history. If you have heart disease in your family, think about a diet that has a bit less fat and more antioxidants. There are similar observations about other conditions like arthritis and cancer.

Generally, if you have a DNA test at this point, it is for one or very few genes. This helps. But remember that most major health conditions are the result of many, many genes acting in concert, not just one gene.

I'll admit it can be confusing with so many different recommendations about diet. To me, this reflects the way in which our genetic diversity makes one diet work well for me and not well for you, while another diet has the exact opposite effect.

4. What is the area of research your company is focusing on now?

Our research focus is to understand how the specific ingredients of diet influence genetic activity. And by diet we include food, supplements, medications, chemicals in water, and even cosmetics, for all of these contain chemicals that influence genes.

The goal of our research is to translate the science into practical day-to-day advice for each person, based on his/her own genetic profile and genetic activity. We want to make genomics something that is useful for each of use every day instead of some industrial science.

This is the same thing that happened when Edison invented the light bulb. Suddenly, electricity was something that every home needed. For us, our success in delivering NutriGenomics to the consumer will make genomics something that everyone will want to use to live their lives a bit better.

5. Where can I refer readers today to learn more?

The amount of literature on the Internet is growing almost daily. Two weeks ago, the commission of the FDA mentioned NutriGenomics in his major speech at Harvard.

We welcome readers to contact us. We are assembling the world's first comprehensive NutriGenomic knowledge base, which we will use to help consumers make better choices. We build our research insights on the actual experience of what real people do.

Our current high priority is a totally new method to prevent viral infections using NutriGenomics. We discovered this in January, and have been working to get government support to conduct this important research. We believe we can develop a way to protect many people from the dangers of certain types of viruses, such as a weaponized flu.

6. Do you work with patients directly or only with their managing physicians or both?

We will work with patients directly and with physicians. When a physician is involved, we will be sure to include the physician in the information loop so the person continues to get the best care.

7. How many genes do you test? Do you prescribe a diet or nutraceuticals based on the results?

In the current stage, we will test about 2,000 genes, mainly for the cardiovascular cluster (cardiovascular disease, diabetes, obesity, hypertension, high cholesterol). The virus testing is planned to be a separate test.

The specific food/supplement recommendations are made directly to the person. These change as the person's genetic activity changes. It would be a mistake to prescribe one type of diet for life. Our genes and bodies don't work that way.

The procedure is simple. We get a sample from you, typically from your cheek lining or blood (if you go to a physician). We identify the genes (genotype) and the amount of genetic activity for each gene (expressitype). We provide you a report summarizing the results. Then, depending on your decision, we will provide you general dietary recommendations based on your genes, or will begin to work with you as often as daily to help you choose what you like to eat.

It is worth noting here that the so-called 'med diet' is based on a month of eating, whereas the USDA model is a daily model. We prefer the monthly approach for the evidence is that eating has a cumulative effect. Another way of saying this is "no one meal will ever hurt you. It is the combination of lots of bad meals that hurts. So knock yourself out."

Fredric D. Abramson, Ph.D., S.M.
President & CEO
AlphaGenics, Inc.
http://www.Alpha-Genics.com

Let's profile another company that helps you eat better by revealing disease risks, so your physician can concentrate on focused prevention and treatment tailored to your individual genetic signature. Today, there are companies such as *Genovations*™.

According to Marketing Communications at the North-Carolina-based Great Smokies Diagnostic Laboratory, the company approaches preventive health care by using genetic analysis to provide individuals and their physicians with critical information to more effectively control present and future health. Genovations uses information stored in each person's genes to reveal disease risks and to help each individual lead a healthier life though focused prevention and individualized treatment.

Genes provide the key to a wealth of information unique to each individual that, once unlocked, can serve as a comprehensive guide to a healthier life. By identifying specific genetic risk factors and changes in dietary, lifestyle, nutritional supplements and medications that are most likely to improve each patient's health, Genovations enables physicians to practice truly individualized preventive medicine.

Most consumers will ask about what happens when you go for a genetic test. It's more than the DNA test you might have taken to look at your mtDNA or Y-chromosome for deep ancestry. Instead, the kind of genetic testing you'd take at Genovation, for example is to evaluate specific portions of your genetic code.

Genovations™ testing evaluates specific portions of the genetic code that vary from person to person. These variations are called single nucleotide polymorphisms, or SNPs (pronounced "snips") for short. Everyone has SNPs—they're what make people different from one another—our hair color, our height, our voice, even key aspects of our personality. SNPs are the very seat of our individuality. SNPs also affect our health.

Some SNPs can make people more susceptible or more resistant to common diseases. Others may make the body respond differently to certain diets or lifestyle habits. Read the *Complete Blood Type Diet Encyclopedia*. What nutritional genomics needs is a complete human genome diet dictionary. With 40,000 or more genes in each person shuffling with each generation, individuals inherit different genotypes.

At the same time, some abnormal genes are inherited and make members of some families more at risk than other families. **Here is an e-letter from Kay Patrick, Product Manager, Genovations.**

From: Kay_Patrick
To: 'Anne Hart'
Sent: Tuesday, July 22, 2003 6:46 AM
Subject: RE: Thanks. Just a few questions

Dear Anne:

I have attempted to answer your questions. Please let me know if you have additional questions.

Why are Genovations tests only available through a physician?

Genomic test results are most meaningful and best understood when they are clinically interpreted within the context of a patient's complete health history. For this reason, genomic testing is best managed by a licensed health care professional.

Genovations tests are available only through trained, licensed health care practitioners. Not all practitioners are adequately trained to interpret and use the genomic information revealed in these tests, so Genovations has taken a leading role in providing genomic medical education to health care practitioners.

What does Genovations ™ testing do?

Genovations testing focuses only on SNPs in the genetic code that are associated with common health conditions, such as heart disease, osteoporosis, allergies and asthma, for which simple treatment strategies exist to reduce risk. Testing also evaluates SNPs that affect how each person's body is likely to respond to specific diets, supplements, and medications.

Therapeutic recommendations based upon the genetic results are provided for the practitioner to develop a treatment protocol. In this way, test results can provide the physician with a "road map" for developing a comprehensive, personalized health action plan for each patient.

Example:

Heart Disease

MYTH: All persons with a family history of heart disease can minimize their cardiovascular risk by reducing their dietary intake of fat and salt, exercising more, and taking a cholesterol lowering medication.

REALITY: Heart disease is not simply a condition caused by excess fat and cholesterol. Research has revealed that there are many other modifiable risk factors, some genetically influenced that can predispose a person to heart disease.

For example, some people have a genetic inability to properly metabolize folic acid in the body. This can lead to a build-up of homocysteine in the bloodstream, causing increased risk of blood clots and atherosclerosis. Yet none of the "one-size-fits-all" conventional therapies listed above would reduce this risk. For a person with this genetic variation, the only way to reduce risk is to take the active form of folic acid, which is not found in common vitamin supplements.

High blood pressure is typically treated by restricting an individual's salt intake. However, not all people have the genetic variation (a single nucleotide polymorphism, or SNP-pronounced "snip") that allows them to respond effectively to a salt-restricted diet. Based on their genetic make-up, they may respond better to aerobic exercise. By testing genetic variations, the physician can better identify which therapy is likely to be the most effective for lowering blood pressure in each patient.

Apo E is a protein in the body that affects cholesterol levels. There are three major genetic variations of the Apo E gene that can affect how each person's body breaks down fat and cholesterol in the diet. These three variations can lead to an increased, average, or decreased risk of heart disease. By knowing an individual's Apo E genetic variation, the physician can prescribe the dietary and lifestyle changes, nutritional supplements, or prescription medications most likely to lower cholesterol levels effectively.

What is the process?

Practitioners incorporate genomic testing into their complete work-up along with other phenotypic diagnostic tests. Blood samples are collected in the office, shipped to our laboratory, tested here in our genomics lab and the reported results are shipped back to the ordering practitioner who schedules a follow-up with the patient.

How much does a Genovations test cost?

The cost of Genovations testing depends on the testing option chosen by the patient and the practitioner. The comprehensive testing program, which assesses health risks for a wide range of conditions (including heart disease, osteoporosis, and immune dysfunction) costs about $1,000-$2,000. Focused testing for health risks associated with a single area or condition costs about $300-$500.

Hope that helps.

Sincerely,
Kay Patrick

Product Manager, Genovations
www.genovations.com

*   *   *   *

An overview of what genes are, how they are inherited, and how many are in the human body is presented on the Web site at: http://www.genovations.com/patient_overview.html.

What Can SNPs Tell You About Your Genotype?

SNPs can even affect whether certain drugs are likely to have severe side effects or not. By evaluating these health-related SNPs, Genovations™ testing allows physicians to develop preventive strategies that are tailored to each patient's unique genetic makeup and health risks. This helps to reduce the guesswork that results from using "one-size-fits-all" approach to preventive medicine.

Genovations™ is a line of tests that is only available for a licensed health care provider (doctor) to order for a patient. The test results are provided directly to the health care provider, who consults with his or her patient. Genovations does not interpret test results for the patient or the health care provider. A patient's health care provider uses the test result information to determine appropriate treatment protocols for the patient. Genovations™ does not provide specific nutriceuticals or supplements recommendations. That is up to the health care provider.

According to marketing communications, "Genovations does not test for SNPs. We currently have 4 profiles that identify specific SNPs related to heart disease, detoxification problems, bone health, and immunology related problems. Each profile identifies between 7 and 15 SNPs." If you want more information about Genovations, write to them, email them, or telephone. Genovations' phone number and email address are listed at its Web site. Since books hang around libraries for years, but phone numbers may or may not change, you can reach Great Smokies Diagnostic Laboratory or Genovations™ as follows: Corporate Headquarters, Great Smokies Diagnostic Laboratory/Genovations™ 63 Zillicoa Street, Asheville, NC 28801, USA

Or go to their Web sites at: http://www.genovations.com/ and/or at: www.gsdl.com. See the Web site at: http://www.gsdl.com/assessments/finddisease/cardiovascular/metabolic_glycemia.html

Read about metabolic glycemia. If you have insulin resistance, or too much insulin pouring out every time you eat certain carbohydrates, you'd want to read more about this and how it changes or affects your arteries and other parts inside your body.

There are excellent sites on the Web to educate yourself before you decide to get tested. Check out your biomarkers. Before you see your doctor, know what questions to ask and what tests to ask for. Then talk it over with your physician.

When you take control over what you eat you are educating yourself as to your body's nutritional requirements to maintain health, stay fit, and prevent or delay chronic illness at any age. Your genetic profile tells you how your genes are responding to the food you eat.

With 30,000, 40,000, or 70,000 genes in each individual and each peson inheriting a different genotype, you'd need a customized diet book for each person on Earth.

Each individual's genetic signature is different. Nobody responds exactly the same way to certain foods. That's why genetic testing helps to *customize* food plans for the individual. Some people have allergies. Others respond to foods gradually and silently inside their arteries and organs.

Another buzz word in the media is 'smart' or 'intelligent' foods. It's not that the food is smart or genetically changed. It's that the food in a relatively unprocessed, natural, clean state is free from vermin, full of life and enzymes. That prescribed food according to the person's gene expression is then healthier to eat for that particular person. Another person could be allergic to it and become sicker.

The food itself is freer of toxins. Additionally, the 'smart' or 'intelligent' food terminology really means the act of tailoring or customizing the food to the person's genetic profile. It does not mean the food is genetically altered and may or may not be dangerous to humans. 'Intelligent' foods refer to the results of your genetic testing.

You're the person being smarter by eating what the results of genetic tests reveal. The results are interpreted by a professional. In addition, you need to find out for yourself how to apply the test results to what happens to you when you eat a particular food combination or take a certain group of nutraceuticals. In making a food such as raw soy milk, for example, you need to know that raw soy can take certain nutrients out of your body if the soy milk is not cooked for a certain length of time before it is bottled, cooled and then consumed.

Blue, purple, and red-colored fruit or vegetables are excellent for your health, if your genotype says so. Eating raw blue berries also will take out certain vitamins from your body unless the berries are frozen or cooked.

You need to know how to process raw vegetables and fruits either by cooking and/or freezing before they are consumed so that eating them day after day in certain amounts won't lead to the leaching out of various vitamins, enzymes, or minerals from your body. Knowledge is important here. When you are prescribed certain foods, make sure you know how to prepare them for the maximum nutrient benefits.

Diets need to be tailored to individuals. There are only four blood types, but individual genotypes are different for each person. You'd need a customized diet book for each person on Earth. Each individual's genetic signature is different. Nobody responds exactly the same way to certain foods. That's why genetic testing helps to customize food plans for the individual. Some people have allergies. Others respond to foods gradually and silently inside their arteries and organs.

Not only do people of different ethnicities react differently to the same doses of medicines, but people of different ethnicities react differently to certain foods. It's touchy to bring in the term "race," but studies have found that people of a certain race react differently to certain drugs for certain conditions, and that people of another race react differently to the same dosage given to another race. You can read in medical journal articles for example, on how African Americans respond to glaucoma drugs compared to Caucasian people. You can read about how the same dose of a drug given to an East Asian and a non-East Asian will be too high a dose for the East Asian.

You can read about foods—how people from Northern Europe tolerate milk better than people from Southern Europe, and how people from Southeast Asia hardly tolerate milk at all in many cases. Then, there is always the exception. People are of mixed ancestry. What you need to learn about is the way intelligent foods work with your genes.

You don't know who will react in which ways at the genetic level to the food or medicine or whether the dosage of the drug or the amount of the food will be tolerated or too high or low. Even with varying dosages of nutraceuticals—supplements such as vitamins, food extracts, and minerals, some people benefit and others show no change. Read the results of conflicting studies in medical articles. If the dosage of vitamins is hotly debated, drug dosages and ethnicity is another topic in the research arena.

Some people have inherited risks for certain chronic diseases. Those people need to eat foods and perhaps take nutritional supplements that will prevent or delay the onset of those problems. Nutritional genomics fills an important need in maintaining health and quality of life. From the ***Institute of Food Research***

(IFR) in Norwich, UK, **Dr. Ruan Elliot, lead IFR scientist,** sent me this e-letter in response to my questions about what is being researched there. Here's the reply.

Dear Anne:

My main research interests are in using so-called functional genomic techniques to define mechanisms by which diet and specific components of the diet promote human health. These powerful techniques are set to revolutionise the way we approach fundamental nutrition research. On top of this, as you will appreciate, there is also the aspect of inter-individual genetic variation, the impact that this has on health and the potential variations in optimal dietary requirements.

To my mind, these two areas are locked together. We need to properly understand the processes by which components of the diet (nutrients, and micronutrients) work individually and together to keep us healthy so as to be able properly to define optimal nutrition for sub-populations or individuals properly based on their genetics.

You can find descriptions of my work and research interests at the following URLs;
http://www.ifr.ac.uk/public/FoodInfoSheets/EDPgenomics.html,
http://bmj.com/cgi/reprint/324/7351/1438.pdf

I hope this is helpful.

Best regards,

Ruan Elliott

For further information, contact the Institute of Food Research, Norwich Research Park, Colney, Norwich NR4 7UA, UK. To view their Web site, phone and fax numbers or email address, on the Internet, go to:
http://www.ifr.ac.uk/about/.

My philosophy about genetic testing is to remember a quote by Richard Feynman, Nobel Laureate: "The best way to predict the future is to invent it." Your genes are hard-wired for certain foods, but not all foods make compatible software. Consumers need a guide book to nutritional genomics. You'll hear terms such as gene expression, genetic signatures, risk, intelligent foods, and tailoring the food to your genotype. What it all means is that your body is looking for customized nourishment.

If you really want to take charge of your own health and nutrition, learn all that you can possibly find out about how to apply the results of your DNA testing, genotype testing, metabolic, blood, ancestry, racial percentages testing, and body chemistry or allergy testing to what you eat, the nutraceuticals and supplements you select, your exercise style and lifestyle. Everything starts at the cellular level, even the way your body reacts to stress and exercise with cortisol or with relaxation. Individuals react to certain types of exercise, foods, or perceived stress situations differently.

Some nutrients, foods, herbal compounds or other supplements cause relaxation in some people and panic attacks in others. Then there are allergies to consider. It starts with an expression of your genes in reaction to the environment. Some people get an increase in ocular pressure from sleeping on pillows containing a stuffing to which they are allergic. One sip of caffeine can start a panic reaction in one person and relaxation in another. Your reaction is in your genes. It's about body type, another genetic expression, whether you have inherited genes for anxiety of certain lengths, and the whole interplay and interface of one team of genes with another group in your body. I refer to this interplay of genes as a "rhumba of rattlesnakes" at the molecular level.

You have a part to play in all this, and it isn't always as the passive patient or recipient. You have the *genomer* and the *genomee*....Be the *genomer* for a change—the person in the driver's seat. Take control. Take charge of how to interpret your DNA tests for risk and diet changes or for ancestry and family history or for any other purpose for which you want to test your genome or any region of your genetic profile. To be able to understand how to read and interpret a DNA test and apply it to foods and supplements and to know how the foods will actually effect your genetic expression—that kind of knowledge is power.

The information is publicly available in medical school libraries, on the Internet's scientific databases, and in various journals in the nutrition and genetics fields. Most of these sources are open to the public without you having to be a scientist to learn about how your body responds to food at the cellular level. Start by joining various online groups, visiting your library, and reading the latest medical and scientific journal articles in the field of nutritional genomics. Network with patient support groups that use genetic testing. Start your own email list message board for consumers who want to learn and listen in addition to sharing resources of information.

The consumer's role is to compare, review, and find out who is best qualified to work with an individual's genes. How does a consumer discern between snake-oil and reputable companies in this growing field? Who is qualified? Consumers need an explanation in plain language what is healthy to eat, not for the world, but for the individual. So it's up to the consumer to do some research and learn a lot more about nutritional genomics.

Beyond food what does an individual's gene expression require in terms of exercise and lifestyle? It's time to educate your body about nutritional genomics and about DNA and ancestry. What foods should you eat and what nutritional supplements (nutraceuticals) would benefit your health? One way to tell is to test your genetic markers. Which genetic markers? The entire genome? Or specific SNPs that signal risk of certain chronic diseases? Nutritional genomics should be available to all consumers, not only those with money to pay for expensive testing. Sure, in the future the price of genetic testing will come down, but senior citizens and parents today want to know what they can eat that will agree with them and their families.

Researching the Web under "nutritional genomics" I found a company in the United Kingdom called Sciona. According to Sciona's Web site at: http://www.sciona.com/coresite/index.asp?p=1, Sciona is a venture capital backed company that researches and develops tests for common variations in genes which affect your individual response to medicines, food and the environment.

There are around 40,000 genes in the human genome. Sciona identifies those genes that influence a certain function such as cardiovascular status and tests for these as a set or 'panel'. This information is then used by appropriately qualified practitioners to provide you with health advice.

Testing of specific genes rather than the entire genome usually is done by various companies at this time. In the future, the number of genes tested may increase, or science may find which particular genes interact with other genes to put you at risk if you eat certain foods, or whether your specific genes work together in such a way that you can eat almost anything without developing chronic diseases.

According to Sciona's Web site testing specific genes for certain chronic illnesses is useful in guiding aspects of the treatment of diseases such as heart disease and osteoporosis, which are influenced by your genes, lifestyle and environment. For the consumer, knowing which foods to eat to influence your gene expression is important.

According to Sciona's Web site, Sciona's team of geneticists, molecular biologists, medical doctors and dieticians work with universities and other companies to identify the significant genes underlying a particular effect. The effects of these, and other factors such as your diet, are then analyzed to give you specific advice on courses of action tailored to your own genetic makeup and circumstances.

Filling out questionnaires allow consumers to think in terms of focus sheets. Besides DNA testing, food choices, and consultations with your physician, you need to focus on your habits and think how realistically you answer a questionnaire. Think about from which direction you want to participate in nutritional genomics—marketing, research, science, consumer awareness, forming support groups, media, or other. What are your basic interests and how can you apply them to this field? You can contact Market America, Inc. at 1302 Pleasant Ridge Road, Greensboro, NC 27409. Their Web site is at:
http://www.marketamerica.com. Here is an e-letter from Andy Aldridge, Public Relations Director, Market America, Inc.

Dear Anne:

After extensive market surveying and testing, Market America and Cellf, a division of Sciona, have partnered together to offer the Nutri-Physical™ Gene SNP DNA Screening Analysis program. This product allows consumers to submit a sample of their DNA to have it analyzed. Once the analysis is complete, consumers receive a report that outlines possible deficiencies along with lifestyle and diet changes that can be carried out to address possible vulnerabilities. To accompany the DNA analysis, Market America also developed a questionnaire that, when analyzed in conjunction with the Gene SNP product, results in a suggested list of customized vitamins and supplements, available in one formula, if desired by the consumer.

The company's Isotonix® Custom Formula makes choosing the correct nutritional supplementation a simple and efficient process. From a Distributor Custom Web Portal, a customer submits answers to a dietary and lifestyle questionnaire and has the option of purchasing a unique custom formula nutraceutical that specifically addresses their individual needs.

Many companies are talking about NutriGenomics. Through our partnership with Cellf for DNA analysis and Garden State Nutritionals for manufacturing and customization, we are actually doing it.

Regards,

Andy Aldridge
Public Relations Director
Market America, Inc.

Consumers Need to Be Involved in Quality Control

You need a voice in quality control. What the consumer needs to understand are the roles of genes in healthcare, and how the *roles of genes* interact when you take in nutrition. Consumers, corporations, venture capitalists the government, taxpayers, and research institutions invest billions of dollars each year to develop this understanding. One example of a consumer group involved in quality control is when parents group together to form their own DNA bank, recruit people to donate DNA for research, and develop databases and Web sites disseminating information on a particular genetic condition.

Consumers can don few or many hats in nutritional genomics. There are avenues to explore varying from watchdog, marketing, research, public relations, parenting, safety, event planning, publishing, gerontology, videography, genealogy, healthcare, to broadcasting.

If you do your research, you'll find that venture capitalists who in the last decade invested heavily in the computer industry are now looking to invest in biotechnology. The power of gene technology drums up business and communication also for patent attorneys, journalists, and inventors.

To participate in nutritional genomics in a variety of capacities as a consumer, you can write to the United States Food & Drug Administration (FDA), Department of Agriculture (USDA), and National Institutes of Health (NIH), as well as university laboratories, pharmaceutical manufacturers, and government agencies worldwide. Get involved in the power of genetic technology at some level. The FDA is the agency that's responsible for 80 percent of the United States food supply, according to a July 1st 2003 speech given by the Commissioner, Food and Drug Administration.

In that July 1, 2003 speech before Harvard School of Public Health, Mark B. McClellan, MD, PhD, Commissioner, Food and Drug Administration, said, "All of you—consumer advocates, representatives of the food industry, nutrition scientists, and other food experts—have a collective commitment to the issues we face at FDA that is integral to our ability us fulfill our mission. And your help is needed more than ever. Now more than ever, we all must work together to find better solutions."

It's important to read this speech and to look at materials on the Web at the Harvard School of Public Health. Click on the Harvard School of Public Health's Web site at: http://www.hsph.harvard.edu/now/jul11/conference.html. Read the important facts there and check out the forums. The current headline at the Web site notes that "A July 2003 conference at the Harvard School of Public

Health 'spurred' dialogue with the nation's food industry on the subject of "'Changing the American Diet' To Improve Health." Read the materials there and think for yourself about how the food you eat is processed and marketed.

In the speech, you'll find key words such as "consumer advocates" and "collective commitment." Your help indeed is needed to work together as a team, to share, with a purpose of finding better solutions. That's why it's important to listen and learn all you can about the future of nutritional genomics. The consumer's involvement is important. The theme emphasized "collaborating to improve the American diet." Collaborating with consumers also means working to decrease obesity and epidemics of diabetes in children.

You can look at the American diet from the point of view of those who work in DNA testing, from those who work with family history and ancestry research, or from those who work in food packaging and processing. It really hits home when you look from the point of view of the consumer or from marketing and product management. Everybody has to eat.

＊   ＊   ＊   ＊

There's another branch of nutritional genomics that instead of only testing your DNA to find out which foods are healthiest for your genes, focuses on *manipulating plant micronutrients* to improve human health. See the article on the Internet, a PDF file on a Web site at: **http://www.ipef.br/melhoramento/ genoma/pdfs/dellapenna99.pdf.**

The volume of imported food is growing each year. Consumers have a field cut out for them—surveillance. As FDA increases its examinations and sampling at borders, consumers can work together to research information about food imports and inspection.

A laboratory can only sample so many products. Consumers can take a role in food security, perhaps looking at industry to identify problems or threats. What the consumer's role entails is better information and collaboration. Everyone needs to keep costs down.

Plant biotechnology of food and feed is another area of consumer interest. If you buy food that comes from overseas, do you ever wonder who oversees the packaging and shipping of those products? Are there really enough inspectors to go around? Consumers worry about the widespread use of sugar in softdrinks. In addition to having your DNA tested, you need to understand how what you eat influences your health at all ages.

Another way consumers can oversee quality control is by forming public interest research groups funded by grant money, private donors, institutions, or the government. You can become a volunteer in nutritional genomics, an ombudsman, a lobbyist, or start your own consumer research interest group.

You can turn a hobby of nutritional genomics or DNA for ancestry and genealogy into a business by affiliating yourself with a university lab which you contract to do testing from your DNA testing clients. There are open doors for consumer involvement depending upon your skills and interests. Nutritional genomics needs public speakers and technical writers to relay to the public what innovations the experts are bringing to healthcare and food systems design.

From running a summer camp for teens interested in nutritional genomics internships or learning experiences to recruiting DNA donors to create a DNA bank or in researching and writing about genomics, there are a variety of doors. Consumers have power in numbers. You can even enter as a venture capitalist with a goal of raising funds even if you have no funds of your own and plenty of determination to learn to ropes.

Don't overlook nutritional genomics for the pet care industry from foods to medicine. Contact the veterinary schools about their research on how foods affect genetic signatures of pets or race horses. Check out the Web site for Research Diets at: http://www.researchdiets.com/.

Research Diets, a New Jersey company since 1984 has formulated more than 6,000 distinct *laboratory* animal diets for research in all areas of biology and related fields at hundreds of pharmaceutical, university, and government laboratories around the world. Nutritional genomics isn't only for humans or laboratory animals. Did you ever think about how your dog or cat could benefit by genetic testing to determine which foods are healthiest?

Talk to your veterinarian to see who is researching how nutraceuticals and better food can help your pet's health, especially when the pet is older. What about nutritional genomics for farm animals or pets? Find out who is doing what kind of nutrition research for better health.

Or you can tap the venture capitalist watering holes and open your own nutritional genomics business. Consumers can hire, outsource, or contract fee-for-service with licensed healthcare staff and geneticists specializing in nutritional genomics research. It's not difficult to raise funds. Billions already are invested in the gene technology industry. Sometimes parents of children with genetic conditions are the first to band together to form DNA banks for medical research. Don't overlook agri-nutrition. It's about what vegetables and fruits you eat that in turn impact your genes.

Medical schools and research institutions often seek federal grant money. Someone is investing billions for this research. Find out who are the investors. Who benefits from this knowledge first? Who has access? Is it the consumer who pays for testing, healthcare, nutraceuticals, and selected foods? It's up to the consumer to take a look at quality-control of the entire nutritional genomics arena.

Everyone is concerned about cell degeneration from a lack of critical nutrition. Unless you know how your genes are impacted by various foods and nutraceuticals, how can you can decide what's good or bad for your body other than looking at family history?

The consumer is concerned with how nutritional genomics as an area of research is applied to healthcare with a goal of disease prevention. For the consumer with no science background seeking a beginner's guide book to the power of gene technology, the first step is to realize that genomics has applications for many more areas of exploration than nutrition, healthcare, archaeology, genealogy, oral history, and population genetics.

So should you get screened? With the enormous variety of diet books on the market, healthy-eating advice abounds. How do you know what works to keep your arteries unclogged and your organs nourished for the combination of genes you inherited? People are different, but similar advice often is given to everyone.

What you need is a personalized report, time capsule, scrap book, or profile that has not only your genetic test results, but everything about your lifestyle, stress reactions, exercise. For example, if your body reacts to exercise by secreting way too much cortisol, should you be taking vitamin C? Will the vitamin restore to a healthier state or further damage your organs? If the vitamin is prescribed, what amounts of it will work best with your genetic expression or genotype? How do you find out whether this regimen is good for you or not? It is to these types of questions consumers want answers.

Information on food groups, vitamins, minerals, extracts, and nutraceuticals in an easy-to-understand format and binder are what consumers need. You also have to learn how your DNA test was interpreted. If someone interprets the test for you, how can you apply the test results to eating smarter foods for your genotype or selecting the right dosages of nutraceuticals? You'd have to keep paying for advice on every aspect of your nutrition as another branch of healthcare.

If you know the language of your genes or can learn it yourself, it makes your road to a healthier self clearer. Consumers need access to and knowledge of quality control. Who is performing the tests? What kind of medical supervision is officiating at the marriage of healthcare to nutritional genomics? Who is at the top overseeing the quality control—the consumer? It should be. You pay to fund the research in tax dollars. So form your consumer groups and look into quality control in the field of nutritional genomics, DNA testing, and related businesses that will open to serve the consumer and the licensed heath-care professional.

As the power of gene technology reaches the masses, all types of offerings will find a way to apply DNA test results to what you eat, how you exercise, what you wear, your prescribed medicines, therapies of various kinds, what music will change your physiological responses based on your genotype, and even what career, lifestyle, mate, childbearing plans, or hobby you choose.

DNA samples usually are obtained by rubbing a brush swab on the inside of your cheek, completing a questionnaire, and mailing the brush to a testing laboratory. Sometimes in the case of a rare genetic disease, blood samples are taken, but most DNA testing is done with a cotton or felt swab or mouthwash. After several weeks, your physician would get a report to interpret for you for nutritional genomics. In testing DNA for ancestry, the report would be sent directly to you.

How specifically is the report customized for you? Even tiny differences in your genes influence the way your body metabolizes foods and excretes toxins. These reports look for *variations*. Your feedback consists of dietary information. Some companies assess your present eating habits and foods with your genetic profile. Advice is offered. For example, some tests look at the type of meat you eat, whether your meat and fish are smoked.

The tests look at the types of vegetables. How many cruciferous vegetables do you eat—such as cabbage, cauliflower, Brussels sprouts, and broccoli? How many raw vegetables do you eat? How many sprouted legumes or whole grains not processed into flakes? Did you know that cauliflower is a low-carbohydarate vegetable, but broccoli is higher in carbohydrates? Did the sprouts you ate have bacteria on them that affected you?

This is important if you are insulin resistant or have too much insulin pouring outing each time you eat a high-carbohydrate vegetable. For example, pectin consumption results in an insulin release. What kind of antioxidants do you eat? How much folate are you taking in, and how does that folate impact your arteries—neutral, good, or worse?

What types of whole grains are you eating? Are you eating your whole oat groats or eating foods high on a high-glycemic index that rapidly turn to sugar in your blood and require more insulin to be secreted which can lead to other problems in excessive amounts.

Does your genetic profile reveal a need for a Syndrome X diet or other measures to halt the excessive insulin secretion each time you eat carbohydrates or proteins? What kind of weight problems or control do you have?

How much sugar do you eat? Do you drink liquid candy such as soda pop? How do diet drinks affect your genetic expression compared to sugared drinks? What kind of saturated fats do you eat and how does your body react to that kind of fat? Do you smoke? What damage has smoking caused your body? What kind of exercise do you get?

Walking moderately may be enough for your body, and too much exercise could bring about high cortisol levels—the stress hormone. Does your body react to exercise as stressful or relaxing? What allergies do you have? All these questions need to be compared against your genetic signature. What genes are assessed by the testing company? Do you have a genetic propensity to alcoholism? Which gene predisposes you to alcoholism or drug addiction when you are under stress? Did you inherit the anxiety gene? If so, what is the right career for you? Do you have the genes that predispose you to panic disorder? Find out how many genes are studied and which genes relate to what physical aspects of your health?

Look for a company with excellent quality control. Recommendations should be practical and proven. Your goal is to improve short and long-term health and prevent chronic diseases or at least delay them as long as possible. If the food you are prescribed don't make you feel better, find out why. The tests you take and the food or nutraceuticals prescribed should be scientifically proven in reputable studies. You can check out the studies in medical journal articles. Make sure they were not flawed studies or studies so old that new studies have proven opposite conclusions.

How can you tell for sure what risks you have and what needs your genes have for certain foods or nutraceuticals without testing *all* the genes in the human genome of an individual? If all the genes aren't tested, how can science really know everything there is to know about your individual gene expression? What's important to know now—going with what you have available? When will the research teams be ready to reach out to the average consumer of nutritional products and healthcare? Today there is a divide between what's going on in research and what's available to consumers. That's understandable because the research is still going on.

Without knowledge of all your genetic markers, how can you develop a plan for maximum health through nourishment? Why must this plan be in the hands of a licensed professional instead of an informed, self-taught consumer? If the consumer is armed with knowledge of what the individual's genes reveal and how it relates to certain foods and supplements, the consumer than can take back control and power over his or her own health.

Knowing your genes is only a beginning. How do you apply that knowledge to practical applications as in how certain foods affect your body chemistry and metabolism? If you have a plan A and a plan B, and you know the unexpected can always kick in, can you still have that feeling of a little more control over the way your body treats you? In a world so out of control, consumers of healthcare are looking for a semblance more of control and power over their sense of well-being and mood.

Diet books written for the masses may not work well with your particular gene expression. Science no longer tells you that you have inherited some gene mutations or defects. So your gene expression causes certain chemicals in your body not to work normally. Instead, you are told that you have an individual gene expression requiring you to eat this instead of that.

If you eat this combination of food in the morning, that excessive insulin pouring out each time you eat high carbohydrate vegetables or fruits won't narrow your arteries so quickly. It sounds so positive. It's practical, and tells you specifically what you should eat and when and how the food affects particular organs or metabolic and chemical reactions in your body as a result of consuming a particular food or nutraceutical. At last, there's a positive solution.

You can use intelligent food to override your defective genes that put you at risk and nourish your gene expression by fulfilling your genetic signature's deepest nutritional needs. How fast are the professional dietitians and nutritionists tuning in to nutritional genomics?

The key for the average consumer without degrees in nutrition or genetics is to learn how to look at a printout of your genetic markers and be able to interpret what that means in terms of which foods and nutraceuticals to consume for your health based on what you are at risk or pre-disposed to come down with given the right interplay of environment, lifestyle, food, attitude, and perception of stress.

It's not as hard as you think to find this information in libraries and in online databases. Any subject that takes money away from professionals making a living giving advice in any field is going to require effort to learn. The key is to find sources willing to share information. You bet they are around, and a lot of scientists are willing to share. Some scientists are concerned about their careers and reputations. They should be. Some are cautious about with whom they share information.

On the other hand, some scientists do not share. It may not be the fault of the scientists, though. If you look in the archives of the Los Angeles Times, you'll perhaps find an article dated July 18, 2003 titled, "Whose DNA Is it, Anyway?" Read that article. It's about a person whose mother has Alzheimer's disease and who, according to the article, was "trying to coax Alzheimer's patients and their families to donate DNA" to a university. It's not only about one person because more than ten thousand people donated DNA for the cause.

How about you starting a nutritional genomics DNA bank, new support groups, non-profit organizations, online discussion groups, or Web sites such as "Moms for Genome-Tailored Meals?" Think about how many consumers, not scientists, whose families have a condition or who are interested in a condition arrange events to encourage people to donate DNA to research.

How many consumers go out to assisted living complexes, senior centers, adult education classes, gerontology workshops, nursing homes and churches or family meetings to recruit people who have relatives with a certain condition such as Alzheimer's? How many consumers reach out to the actual Alzheimer's sufferers to recruit people to give DNA to science—to research, often located at or connected to various university laboratories? This is one way consumers get involved in science without necessarily having credentials in science.

All it takes is a deep interest to take action and learn more about the subject. Some research companies are independent, founded by scientists, but work closely with a variety of universities. Consumers are involved every day in science, usually on the end of recruiting people for donating DNA, planning events, fund-raising, writing about in the mass media, encouraging children to take an interest in science, acting as ombudsmen in nursing homes, running consumer awareness classes for seniors or parents, working with genealogy societies for various ethnic groups whose DNA is being studied by scientists for diversity, geography, or genetic diseases, and philanthropy are but a few ways to get involved in scientific research.

According to the article, you have a situation where DNA was used to search for Alzheimer's genes. Science itself has a great mission, to someday find a cure. The DNA is like a pointer. What the consumer is at risk for is that the research will stop. When the research stop for whatever reason, who controls the DNA you've donated? Read the Los Angeles Times article, "Whose DNA Is it, Anyway?" Think about your own DNA and thousands of others who have donated for testing. Who controls the DNA archive when the research is halted in mid-stream?

Who owns all the DNA samples—including yours? Who will inherit and use the samples and for what type of research? What if you have a group of family members suffering from a condition that inheritable and you don't know whether you'll get the disease or not?

What if you donate DNA along with others hoping it will point toward a cure, but the project ends? It's like the old question of who owns the living embryos in vitro when the project is over? Almost anyone with a nursing, social work, or other allied healthcare background can open a home-based or other business part time online or fulltime recruiting people for medical trials.

Around the world, people give their DNA for medical research. Why do people donate DNA? It's not always to find their ancestry or ethnic origins, because in many cases, results become statistical and no feedback is given to any individual or even to members of an entire group. What shows up are anonymous statistics in scientific journal articles or books. People often give DNA in the hope that science will find a cure for their or their family member's genetic disorder or inheritable disease.

People donate DNA to see whether a new test will work better for them, such as a test of racial percentages, or in the case of a disease, a test that will reveal risk that could be overcome with certain foods or medicines or even gene therapy. People hope that the medical field has a way of injecting normal genes into a child or person that will take hold and correct a defect in the existing gene expression. Part of gene therapy is about the introduction of new genes that could fix the old genetic problem, and often it works in certain cases.

What consumers can do for themselves is to take part in creating lending libraries for DNA. You need to have lending libraries open not only for scientists, but for the consumer to at least read about research. Of course, scientists who actually work in laboratories with DNA could enjoy the free and open access to the DNA managed by special DNA librarians. It helps research move forward. As a consumer you are a taxpayer.

Do you know your tax dollars are paying the cost of much of this scientific research, especially the research going on at state universities? If you are paying for this research, then the scientists are public servants to you, the consumer. You have the right to this information as much as any scientist, since you are paying the bills for the research.

The problem is that not all scientists freely share DNA all the time hoping some other scientist will find a breakthrough. It reminds me of competing journalists on different papers vying for the news scoop of the day. The DNA is collected and sometimes not shared at all.

The reason is that a scientist's career is at stake. In research, a scientist's reputation and job depends too often upon breakthroughs that help the career. Will sharing lift a scientist's career by his or her bootstraps? Sharing could mean the other guy finds the breakthrough first. Then in walk the patent attorneys. Scientists can file patents on genes.

What if your child is autistic? What scientists will share DNA with other scientists in research on autism or Alzheimer's or any other condition? What's your role as a consumer? It's to create a situation for cooperation. Consumers tend to form patient support groups and to build their own DNA banks. That's only one way of taking consumer action in a positive way to move research forward and point toward solutions and cures for the benefit of the health-care recipient—you, the consumer.

What can you do with nutritional genomics that is in your interest? You can continue to build DNA banks backed by consumers with an interest in nutritional genetics or DNA ancestry, or gene testing for disease markers. Share among yourselves. After all, you are not all scientists, and therefore have no career in science about which to be concerned. I speak as a consumer here rather than as a science communicator.

What you can start as a consumer is a movement to pool DNA samples into a managed library and DNA bank open to all scientists. Consumers could have access to reading about the research and learning to interpret their own DNA and other genetic tests of SNPs, since most consumers don't have access to laboratories to perform research. The time for the consumer's role in moving research forward is here, and consumers must learn all they can about how to interpret the entire human genome.

Your first step would be to make a list of the various institutes that study particular conditions of interest to you or written in your genes as your risk condition. Start with the ones funded by you as a taxpayer. Learn more about what your genes show you. When you talk to scientists, you'll find out how many will or will not share anything from information to DNA. Talk to the scientists. Get on emailing lists where scientists talk among themselves. If it really were true that scientists never share information, there wouldn't be any libraries or databases. Once anything is published and is not classified by the government as secret, it's shared through journals and libraries. The idea is to do the same with the raw material, the DNA through gene banks and DNA libraries as well as databases.

You already know some scientists don't like to share raw materials such as DNA with other scientists. Just ask most biology educators at universities to share their experiences with geneticists with consumers, let alone the media. If you think your DNA is a gift to the public like your donation dollars, think patents.

What if you're a consumer who works with donors? What if your dream is to see children with certain diseases tested for little cost?

What if there are battles between scientists and hospitals for patents? What if a hospital takes out a patent on your DNA or those of a bunch of donors you as a consumer recruited from your efforts, support group, or fund-raising events? What if the hospital is awarded the patent and starts to charge fees?

What if that results in some testing programs closing? It happens over and over. What if the costs of developing certain types of DNA or other genetic tests are or remain too high? These are all the questions that consumers want answered by scientists, hospitals, research institutions, universities, and licensed healthcare professionals.

When a consumer so enthusiastic about learning all about genetics, DNA for ancestry, or nutritional genomics put effort into raising money and providing DNA donors, the consumer wants to be sure that what's used for research doesn't always go to pay for patents that result in fees and costs to the consumer or anyone else becoming unaffordable just to create or invent the new genetic tests.

Media has created videos romanticizing "gene hunters" poised on the cutting-edge of a new frontier—inner space. Gene hunting is entertainment. Yet there are little controls or rules—not even a law that requires scientists to share DNA with other scientists. If gene hunting is another Wild West frontier focused on individuality of gene expression, where are the public interest research groups here?

Back in 1975 I was a book author employed in a temporary job full time for ten months writing the consumer manuals and handling all incoming calls to the consumer complaint center of a public interest research group. My job was to write a book that ended up in the Attorney General's office. The book was a consumer manual of how to effectively complain.

Today, the consumer needs a guidebook or manual to learn how to effectively organize what should be in the public interest—research. That research could be anything from the frontiers of nutritional genomics to the study of food on a disease or the disease itself. Sometimes research is shared with the public long after everyone else knows about it. If it's archaeology, it's shared quickly. If it's gene research, that's another story. Sometimes patents get in the way, sometimes costs, and sometimes egos.

Databases are open to the public. The bioinformatics profession manages databases of bioscience information stored in computer software. You can get a certificate in bioinformatics at many community colleges or through extended study programs at various colleges and get your foot in the door of bioinformatics through an internship course.

On the other hand, you have scientists, universities, private companies, and research institutions vying for money and/or information being returned to them for their work in giving information to the databases that hold DNA information. It's all about your taxpayer dollars.

As consumers, you pay your taxes to the government. In return, the government takes your tax dollars and dangles that big carrot of federal grant money in front of the institutions employing the gene hunters. And if those scientists refuse to share DNA with other, competing scientists, well, they won't get the federal grant money that's really your tax dollars.

So who's at the root here who should be in control? It's you, the consumer because you supply the government with your tax money that makes up the federal grant money that goes to the many of the places that employ scientists who test DNA for research.

Write to the National Institutes of Health and ask who is sharing what with whom. How are your questions answered? You could get yourself a career as a grants writer and work your way up to become a grants administrator. That's a long career route where you never know whether or not you'll be chosen for the job. The highest you might get is as a serious student of grants administration research. Another route is to ask scientists why sharing DNA could destroy their careers. It's all a matter of who gets to publish first.

In my quest to find people to chat with, a few scientists turned me down not because I don't have a science degree, and not because I don't work for a major newspaper or magazine, but because I'm *the first journalist* to write a book on genomic nutrition before the scientists have had a crack at it. And the book isn't for scientists speaking to other scientists, but a consumer's guide to nutritional genomics and ancestry by DNA testing.

Consumers don't need to be slapped in the face with technical jargon unless the genetics terminology is defined in a glossary. The purpose of this book is to inspire consumers to look into the subject of DNA testing not only for deep ancestry, molecular genealogy, or archaeogenetics, but for tailoring intelligent foods to their genetic signatures. If the buzz words in the news are nutritional genomics, then now is the time to organize. The entire area of DNA testing needs direction, quality control, and applications to what consumers perceive as critical needs.

You see dozens of books published on population genetics, archaeogenetics, the peopling of various continents, mitochondrial Eve, ethnic DNA, diseases and

DNA, ethnic genetic diversity, food intolerance, allergies, inherited illnesses, aging, and DNA for ancestry. How many general-reader type books are published on smart foods based on genetic signatures and SNPs rather than chemical, blood, or metabolic signatures? So positioning yourself first matters if you're a scientist who must publish and patent. If you're an educator with an academic institution, positioning yourself first in publishing is crucial to your career, even factoring in getting tenure.

A consumer's world is based on sharing. A scientist's world is based on fear of competition. It should not be that way, but it is. And few will admit it. What competition? There's no scarcity here. There's room for everyone? Then how come so many excellent PhDs in various sciences can't find jobs in their field?

If you want to check this out, talk to recruiters in the sciences and find out how many people are out of work with PhDs in a variety of sciences. Is it age discrimination? Ask the recruiters for their experiences. They will share with you in most cases. Try writing a book on careers in genetics. See what the recruiters have to say to you. Some scientists at some research institutions also will tell you that before they can share DNA, they have to get consent from everyone who donated DNA.

What do you do as a consumer? You don't work alone. Power is in numbers. You organize, get people together and pitch in as a growing group to start your own DNA bank, perhaps consisting of DNA donors. Then you'll have the problem of contacting universities and researchers at a wide variety of places to see whether they are interested.

For the average consumer who doesn't want to take any action, just find out what foods are healthier, the problem is simple. Just research your own genes yourself. All your genes interact as a team. And as all these genes work together, you usually can't point to one gene as being responsible for one action or disease. You can have a defect in one gene that causes a problem or only a risk.

Both consumers and geneticists are interested in finding out what mutation happens in a gene that leads to chronic illness at any age. When a gene oozes really bad protein or the wrong amount of protein—ah, oh.

The interaction between the rest of your genes and the chemical and metabolic systems go wild at the cellular level. You get sick. Gene hunters who study the mutations in your genes can study mutations for ancestry or for sickness. Different genes are studied for ancestry, and muations there have little to do with illness. So where does the consumer go to start? That depends on what you want from your DNA—ancestry, or a prescription for intelligent food?

If you're talking about illness, you start with patient support groups. For ancestry, it's the genealogy groups, ethnic groups, and DNA mailing lists online or start an archaeogenetics club or e-mailing list. You can start your own academic journal. For nutritional genomics, the field is wide open for you to start your own DNA collection banks. Who controls your DNA—you or the research 'industry'? You, of course. You control the direction of any research performed on your DNA, and it's time to take DNA by the reigns and direct it—yes, you the consumer. Your tax dollars fund research. So take control of your own DNA.

In the field of nutritional genomics, you, the consumer must take charge of your DNA and begin to build collections in DNA banks where the DNA donors are in control of the direction of the research. Nutritional genomics is not only for parents of children with genetic defects that need the research so urgently. It's also for anyone who wants to know what foods to eat to stay healthier for longer periods of time.

You can start organizing family networks to review and compare the nutritional genomics industry. You can build collections of DNA in your own DNA banks and create databases and DNA libraries. Instead of re-inventing the wheel by duplicating the existing databases, fill the gaps where your efforts are needed and rewarded. Go to a variety of genetics, foods, and DNA-related conferences and let people know you are starting your own nutritional genomics DNA bank and/or intelligent food bank.

The skill you need is not a PhD in genetics, but the desire to persuade people to come together. You need to be a catalyst. If you're one of those unemployed people with science training and are good at public speaking, organizing people, and fund-raising, it could be the career you're waiting for.

If you are a parent with no particular skill other than public speaking and an interest to learn all you can about nutritional genomics from self-education, it's a way to take action. Be prepared to outwit Mr. or Ms. Medical Research Politics just as you have outwitted corporate politics and the games your mother never taught you. If you are lucky enough to be a homemaker, this is the perfect part time career—finding loopholes in medical research politics. It works best if you are not a doctor's wife. Being a doctor's mother is even better for this career. And if you're a patent attorney or work for one, this is another ball game.

The first loophole is to form your own group. Work as a group and put your power in the numbers. Work with scientists in a group also. If you have DNA to donate, give it to a group of scientists that you can organize as a group to work together. There's the problem of individual scientists sharing DNA with one another. If there's no group of scientists, ask them to form a group to work with your group. Organize for nutritional genomics.

Pool the scientists, not the DNA. Otherwise, you'll have a bunch of DNA on your hands and no place to store it. You've got to organize scientists together in a group to work with your organization. First, you need to start an organization. It doesn't take a lot of money. It's about people working together. Consumers aren't afraid of competition. Scientists are afraid of competition because positioning oneself first is the rule of the game if they are to keep their jobs or get ahead.

Consumers are are parents, kids, or retirees, not scientists or medical researchers whose careers will be harmed by grouping together to share information or DNA. Most scientists will not share DNA with other scientists. You can go to scientists in other countries and talk to them. See if they will pool and share DNA for research. There's a far better solution. Start your own DNA bank. You organize parents—especially ones with a little money or prestige in their non-science careers. These elite parents have enough smarts to do fund-raising and hire people to collect the DNA to put in your own DNA bank.

What kind of people perform the best fund-raising and organizing? Public relations people—extroverts, people-people, public speakers, entertainers, and the media, for starters. Don't bother the introverted scientists who need to hold their jobs by positioning themselves first in publishing the fruits of their research. Contact public relations and marketing people such as fund-raisers and marketing communications managers.

Hire people not afraid to speak out and not afraid to learn about nutritional genomics. Start a DNA bank, a nutritional genomics resource center made up of many families. Collect a lot of DNA. Ask for private donations. Share your DNA bank's DNA with any scientist. Let the medical schools know about you. Again, you don't need any type of academic degree or credentials to organize a DNA bank. You just need to hire a staff to collect the DNA and make it available to medical schools and any other scientist for research. Who owns the DNA? The donors.

If you want to network with other parents and consumers who have done this, contact people who have organized their own DNA banks, such as the Autism Genetic Resource Exchange. They are on the Web at: http://www.agre.org/.

Talk to the medical schools that use the DNA banks. Read the Los Angeles Times article, "Whose DNA Is It, Anyway?" July 18th 2003. The article also mentions how the Autism Genetic Resource Exchange got started. You can do the same for researching nutritional genomics, a field that got started only back in 2000, and you don't have to have a background in any science to organize people, hire staff, and get started recruiting families to join you in creating your own DNA bank. Do you know how much you are needed by the medical schools?

When you speak to scientists, you may get cooperation, but you'll also come across someone who will tell you to scram because you don't have credentials, or in my case, because I make up stories for a living as a novelist, overlooking 35 of my non-fiction books. Don't let anyone distract your attention from your goal by trying to focus your attention on perceived shortcomings. Anyone who turns you away probably wants first position. True love is about making sure the loved one is positioned first. What nutritional genomics needs is a little tender loving care.

Who do you contact as your first customer? Try the *research-oriented* medical schools. They don't have anywhere near the resources to collect as much DNA as they need. Contact the heads of the molecular genetics departments at the medical schools.

Not everyone interested in nutritional genomics will want to start a DNA bank devoted exclusively to nutritional genomics. There's a place for all levels of participation. You can hire a DNA collector or learn how to collect DNA from people if you're a people-person. You can raise money from donations or ask your state for money. Have each donor sign a statement allowing the DNA to be shared by other research scientists.

Find someone who is associated with a reputable laboratory or with a laboratory in a medical college to store the DNA samples. Talk to biostatistics professionals. The more DNA you collect, the better it is for the numbers crunchers. Not much is learned from only a few DNA samples unless you're comparing Neanderthals to modern humans.

Nutritional genomics as an industry and area of science needs its own support groups. Most diseases have support groups. It's time support groups were formed for preventing or delaying genetic risks some people are pre-disposed to from occurring through learning more about smart foods tailored to individual genetic signatures.

What's your solution? Since scientific research is for the benefit of the consumer of healthcare and nutrition, start by going to other consumers and getting feedback by reviewing, comparing, and polling consumers on their experiences with testing. Are the consumers satisfied? What would they like to see improved? Today's consumers of genetic testing have enough money to spend on tests that help them make decisions about food, health, or ancestry research. You've seen Michael Moore's TV Nation. How about creating a video called Nutritional Genomics Nation? If not a video or a media presentation, how about a database and a Web site?

Still feeling a bit out of control? No need to. Just talk to the professionals in the field and make friends. Aside from professional competition that makes some scientists want to position themselves first in publishing and patenting, scientists really are friendly, welcoming people—especially when you are positioned to give them the kind of publicity in the media they want at this time. Nutritional genomics is a buzz word in the media. It's a hot topic this year. Scientists are finding this one-chance shot at getting positioned first in the media by publications of the highest repute.

Librarians at medical school libraries and publishers really like it when you read magazines that would otherwise be read by only a few. I used to spend months reading journals at the UCSD medical library to get ideas for topics that would result in books—either novels or nonfiction how-to series. So you, the consumer, also can learn the important connection between how to read the results of a genetic test and how to apply it to choosing the right food.

Right now, that key is in the hands of licensed healthcare professionals and scientists. It's time the consumer learned how to interpret the results of genetic testing and how to apply the results to a better way of eating. Self-education is the answer. And even if you put your life in your doctor's hands, how do you know whether he or she is qualified to make the connection between your genes and recommended foods or whether those foods will work with your specific gene expression?

Find out what shows up in your relatives and what you may or may not have inherited at the molecular level. A test of your biomarkers will at least give you that handle on your own healthcare, foods, and lifestyle. The point is to educate yourself about your individual genetic signature regarding risk and the implications genetic markers have on your health now and in your own future.

Keep your profile private, and use the information to choose the best possible combinations of foods to help prevent or delay any chronic illness for which you may be at risk. You don't have to give your genetic profile to anybody who could use it against you such as insurance companies and employers. What you can use your profile for is to choose the foods that your body needs. Let your gene expression show you what to eat, how much, and when.

When you look for a diagnostic lab, keep in mind that you want to deal with companies that provide their testing to physicians or similar licensed healthcare providers and not to anyone from the general public. The reason is that you need to know how to interpret these tests and how to apply what you learned.

For example, if genetic testing reveals your at risk for a certain chronic disease, how would you know that unless you can look at the genetic markers, see the risk, and be able to judge which foods and supplements would cut that risk. What I'd like to see in the future is that the consumer would directly be able to get that information from publicly available sources while maintaining privacy.

What I like about the company is that they have a Web site with clinician support where you can educate yourself about what to look out for in your own body so you can at least ask important questions when you see your doctor for testing. And for physicians and clinicians, it's an excellent site to become more informed in specific areas.

The ideal situation would be to see the results of your testing and be told what risks you have. Then you'd be able to buy a book or look up enough medical journal articles. In the future, as a consumer I would like to see more consumer education so that your preventive care wouldn't always have to be solely in the hands of your managing physician.

You need to have more control over what you eat to lessen your disease risks. For you as a consumer to concentrate on focused prevention and treatment tailored to your individual genetic signature, you need knowledge. If knowledge is power, than those with the knowledge you need and don't have leaves you pretty powerless. So you have to educate yourself via books and medical journal articles, the Internet's Web sites and asking questions.

When you see your doctor, often you have only a few minutes to ask questions as doctors are pressed for time. Make a list of questions you want to ask. You might want to compare opinions and answers between alternative health care licensed professionals and your usual primary care physicians or HMO nurse practitioners.

Then check out with medical journal articles and news or even consumer feedback and reviews of what you're seeking. If you don't see enough Web sites that review genetic testing companies or research institutions and if you don't see reviews of biotech companies doing research, start your own Web site to review the many companies springing up.

Compare the nutritional genomics companies that deal with licensed healthcare professionals, and get answers, opinions, reviews, comparisons, feedback, polls, and consumer reports. You have consumer reports on cars, mattresses, and washing machines, why not a consumer report site or publication on genetic profiling for nutrition and health? Start a public interest research group on companies that look at genes or look at disease in various ways. You'll learn a lot from consumers just as teachers learn more from their students than from many of their books. Ask for feedback. It's your genes.

Learn the terminology. Find out how nutraceuticals affect your body before you decide what nutraceuticals to take—vitamins, minerals, antioxidants. What works for people with your risk, may not work with your specific genetic signature.

What foods can you as an individual eat to lessen the risk or delay the onset of the chronic ailment for which you are at risk? I'd like to see the entire genome tested, not only a few SNPs for the major chronic diseases, but all the genes.

Since cracking the human genetic code in its entirety is so new, science may not have the full impact of which genes react with what foods to build a healthier you. Does anybody really know today which genes are responsible for the way your body reacts to certain foods, medicine, or lifestyles?

Science can tell you the specific SNPs for certain chronic, degenerative diseases. Think about it, is that all there is—as far as all the genes responsible for those chronic diseases? What about the rest of your genetic signature? Are there hidden files at the cellular level?

Intelligent foods are attracting the attention of the big, international food design systems firms. Who else is interested in nutritional genomics? Rushing into the buzz about nutritional genomics include patent attorneys, genetics counselors, universities, pharmaceutical manufacturers, and physicians. Nutritional genomics also attracts the interest of naturopaths, the alternative health and health food industry, and merchants of nutraceuticals. Then there are the big, international food systems design companies. Anybody interested in health and healing or food and nutrition is paying attention to research in nutritional genomics. It's time for the consumer to do some research.

What is your stance on nutritional genomics? I visualize peering inside my genes every morning to know how my body responds to carbohydrates. I can feel too much insulin being released, hitting me like a bomb when I used to eat donuts and coffee the first thing in the morning. A few minutes later I was in tremors, caught between the sugar and caffeine. Needless to say, the whole process led to hardening of my arteries and more.

Only when I switched to a diet that calmed my body type, intelligent foods for my biomarkers, did I realize the connection between what I ate, how I exercised, and what changes were happening in my body in the past sixty-plus years. What I needed was foods tailored to my genotype. My relatives eating processed carbohydrates and trans-fatty acids didn't last very long. As the only survivor, I had to listen to what my body was communicating to me at the molecular level, to be aware. The best way to be aware is to listen to how your genes respond to what you eat.

Genes communicate beyond a chemical and metabolic level of consciousness through the language of biomarkers. They alert you to risk or no risk in a measurable, physiological way. And the language shows up not as words but as risks on genetic tests. Sometimes genes mutate or make mistakes in copying. There also are other reasons for defects in specific genes. Or you may not have any defects. Your biomarkers evolve over millenniums based on what your ancestors ate and the climate in which they lived for the past ten, twenty, or forty thousand years.

You can do something about risk such as lower your homocysteine levels with nutraceuticals and vitamins if the levels are too high. Look for clues in family history, your own blood tests and physical exams…and most important, your genetic risk profile.

That's why knowledge of healing foods tailored for your specific genetic profile should be your right. There are food and nutrient solutions to most risks that show up on tests. It's better to know how to overcome the obstacle than to remain in the dark about what foods will delay or prevent future or present chronic illness. Information about your genetic profile should always be private.

When should it not be private? What if you have a contagious disease or apply for a commercial pilot's license or drive trucks or busses, trains or boats? Then regular medical exams will reveal the expression of your genes as you age. What is your opinion on privacy? Have you looked at the polls on the subject of nutritional genomics? What do you think about genetic testing for nutrition and nutraceuticals? There are allied fields to explore as a consumer.

Research the publications and articles online or in print on phenomics—the science of tailoring your medicine and healthcare to your genetic profile. Explore proteomics (drug discovery). Read about bioinformatics. The information is either on the Web or in medical and science journals you, the consumer can find on the shelves of university and medical school libraries.

You can purchase a library card and make use of medical and science libraries at universities. As a senior citizen in lifelong learning, I was able to purchase a library card through a retirement-age group at my nearest university. There are medical journals you can read in the library or subscribe to. Take notes. Use the Internet to read medical and scientific journals online. Use the library photocopy machine for articles you want to take home if you aren't online. Do your homework and teach yourself about the wonders that are out there in the burgenoning field of nutritional genomics.

It's true that a little knowledge can do a lot of harm, but what you're interested in is for the consumer to be able to self-educate at least to the level of the media, and to have the same access as the media to see what is evolving in fields that look at disease in new ways and in looking at health and preventive nutrition in new ways.

For example, when I used green tea extract to cure my gum problems, my dentist was happy for me. Right now nutritional genomics has buzz appeal. It's in the news. At the root is change. The way scientists look at your health today is that nutrients play your body like a violin. Your entire system is an orchestra, and you have to listen to the music.

How does an orchestra of musicians play together in sync? By being interactive. Your body is part a team. And the systems biology that nutritional genomics is all about works interactively. First science looked at the parts, and then the whole person. Now science looks at the orchestra playing together in sync, the systems biology at the genetic level. It all works together, and what you eat may be able to keep the symphony in sync, playing a healthier, more stable rhythm. Nutritional genomics is more about your individual genes that your ethnic group as a whole entity. What if you didn't inherit the gene to digest milk properly?

In the book **Archaeogenetics**, McDonald Research Institute Monographs, 2000, there's an excellent article based on a study of lactase diversity titled, "*Lactase Haplotype Diversity in the Old World*," (chapter 36) page 305, by Edward J. Hollox, Mark Poulter and Dalls M. Swallow. Science knows that lactase persistence is a genetic trait. It shows up in various frequencies in different populations.

The point is you can't point to genetic markers to distinguish one ethnic group or race from another in order to prescribe a certain dosage of a drug or food because there is diversity in individuals. You have to look at the genes, the genetic profile first to see whether the individual inherited a specific genetic marker.

On the other hand, some members of some races react differently to some drugs and foods, and you can't prescribe for one person based on how the majority of his or her ethnic group or race reacts to the drug or food without seeing the genetic profile to see what the individual inherited. I say before you prescribe a dosage consider how the person's ethnic group reacts to that dosage....but know what genes the individual inherited first. That's at the root of nutritional genetics. Because it's such a hot buzz word in the news, there is lot of discussion about the changes going on.

Scientists are becoming mighty particular about the reputation of who they open up to because of the old adage that says your own scientific reputation in a relatively new field in the throes of change depends upon not only on who you talk to, with whom you're friends, but also on the reputation of the publication for whom the media person works that you open up to.

That's why it's nearly impossible for a freelance book author who works for no publication and is not under assignment by any editor or publisher to get an interview with a scientist willing to talk about nutritional genomics. I found *almost*, but not all doors closed as far as getting scientists to chat with me online about what's new and what's news in nutritional genomics. Yet, as you see here, some excellent scientists in the field did give me information at no cost to me for my book on their wonderful, new and changing field.

On the other hand, scientists working with DNA testing for ancestry or archaeogenetics, including those in Europe, greeted me warmly. They readily provided comments and quotes freely at no cost to me and with the friendliest of attitudes. So for those scientists who treated me equally well as any other media professional, thank you, I deserved that. I don't write for the tabloids. I don't put down anyone, ever. I write books for the general reader about choices.

I read scientific and medical journals and news and go to conventions and seminars with other media people writing about the sciences and belong to professional associations for journalists and authors who write about science. The scientists mentioned in this book from the nutritional genomics industry were most courteous and responded quickly to my email. They are friendly people who gave me the facts I needed for this book, responding immediately to my email. For busy people, that is awesome. Thank you all again.

Scientists need to protect their careers and reputations, but they also need to remember that the consumer is the bread and butter of the nutritional genomics industry. What the consumer wants most from the nutritional genomics industry besides the profile and the prescribed foods and nutraceuticals are respect and privacy. Grandma knows best here. It's not a matter of preaching to the choir, of scientists selling to physicians, licensed health care professionals, or genetics counselors. The consumer's body as a recipient of healthcare is involved in the product, and the product is more than a genetic profile in a database. The key word is choice.

What's abuzz about nutritional genomics is that one size doesn't fill all ethnic groups when it comes to prescribing dosages of medicine or certain foods based on one's race or country of origin. For example, in the past, many people who lived in certain parts of northern or north central Europe for thousands of years mutated a gene to digest milk without symptoms such as gas, bloating, diarrhea, and cramps.

Many people who lived for thousands of years in Southeast Asia did not inherit a mutation to digest milk without symptoms. You can't assume the ethnic group reacts as a whole because there is diversity. Not all Northern Europeans can digest the lactose in milk, and not all Southeast Asians get the runs and gas from drinking milk because of lactase intolerance. That's just one example of

That's why nutritional genomics speaks out in the same way that archaeogenetics by DNA testing speaks about deep ancestry and population expansions. My all-vegetarian diet of soup and salad followed by a lot of muffins and frozen yogurt desserts eaten in my forties and fifties was making my insulin resistance/ hyperinsulinism worse. The burgers and fries I ate up to the end of my thirties didn't help. Midway into my sixties decade I decided to eat smart foods, but which foods were right for me?

I've heard while chatting with people interested in DNA and nutrition, phrases such as "*eat 'bright' for your genotype.*" It seems to be a word play on a NY Times best-selling book that I've read a few years ago titled *Eat Right 4 Your Type* by Dr. Peter D'Adamo. The Web site is at: http://www.dadamo.com/. There are lists of medical journal articles in the book, and I was able to read the medical articles referenced explaining the scientific basis of how lectins work. I followed the Blood Type Diet. I enjoyed the Web site designed to educate anyone about the scientific links between blood types and nutrition. Dr. Peter D'Adamo, is the author of *Eat Right 4 Your Type, Cook Right 4 Your Type*, and the *Complete Blood Type Diet Encyclopedia.*

The popularity of these diets recommended for the various blood types—O, A, AB, and B are backed by medical journal articles of scientific studies referenced. Eating according to your blood type is a theory, but backed by scientific studies. The entire concept of genetically individualized nutrition seems to have evolved starting with blood type. For example, there are scientific studies of how various lectins agglutinate the blood depending on one's blood type. You can see reviews of many diets at the Diet Reviews and Information Web site at: http://www.chasefreedom.com/eatrightforyourtype2.html.

It's as if the sciences of genetics and nutrition had its roots beginning with blood testing regarding how certain foods affect people with certain blood types. Before the entire human genome code was known, scientists used a metabolic and chemical approach to study how one's food intake influenced one's health. Blood was tested, physiological responses, metabolism, glucose, and other tests to measure anything from allergies to whether one had diabetes. Today we have the whole genome at our fingertips. Tests are costly, but the science is here now.

We don't have to wait another ten years to know how our genes react to food, exercise, environment, relationships, stress, or lifestyle. For example, if you look on the Web you may come across something like this: some people who prescribe nutraceuticals based on blood type might suggest that if you have blood type A, and your body is full of cortisol after heavy exercise, change the exercise or take some vitamin C. Is this medical education or medical advice? It's a thin line. And the person who puts this suggestion on the Web may or may not have an M.D. Naturopaths and nutritionists can make suggestions on the Web. How should you react?

You can ask your own physician when it comes to medical advice. On the other hand, is your physician trained in nutrition or nutritional genomics? Will you be referred to another specialist? Would that specialist be a nutritional genomics consultant, a naturopath, a genetics counselor, a nutritionist, a nurse practitioner, a nutraceuticals company, a testing lab, or who? That's why the consumer needs to take control of his own research about healthcare.

Scientists look at proteins in food and allergies, or whether a certain blood type became agglutinated when a certain food was consumed. Back in the era of World War One blood testing was linked to population genetics. Today, research on the entire human genome code has revealed a more comprehensive way of studying health and nutrition through nutritional genomics. Instead of only looking at blood type and the way food affects you—because your blood type is reflected in how every cell in your body reacts to a certain food—science looks at all your genes.

If you're a consumer with no science background, how can you make money in nutritional genomics? You can put up an informational Web site. Compare, review, and evaluate the genetic testing companies that have contact with consumers. You can compare, review, and evaluate the research companies and the university programs. Basically, you don't need a degree to develop a consumer-oriented information Web site and/or database or an annual book of facts published for consumers. What you can do is compare what's available to consumers in the field of genetic and genome testing. It can include nutrition, ancestry, pharmacogenetics, nutraceuticals, and related sciences such as phenomics.

You can provide services in a variety of categories even with no science background to start with by listing information on companies as a type of consumer report on the nutritional genomics industry, on the DNA testing for ancestry industry, and related health and nutrition or nutraceutical firms. If you have no money, begin by asking the reputable companies you list for funding.

Talk to experts who know which companies are reputable who can be asked to fund your online business. Ask the food industry for funding. Talk to the patent attorneys. If the universities have little money for research and can't help you, talk to those where the universities go for funding. What government agencies do the universities contact for funding their research in nutritional genomics? What large corporations? Which philanthropists? Start with the largest, international food companies and the healthcare industries.

Then work down to the nutrition companies and the people in alternative health care. If you only want to give out information, that's helpful to the consumer. Develop a database. You want the companies to sponsor your efforts. The first step is to write up a plan and keep knocking on doors. Do a bit of fund raising. If nothing happens, create an informational Web site anyway. I did at www.newswriting.net. My purpose was to give information on genetics, broadcast my talks, and promote my books with excerpts and articles.

Yours may be to compare and review companies involved in any aspect of the DNA testing or genomic profiling arena and possibly to review books and publications. You could include feedback from customers of DNA and genome profiling and testing firms, research organizations and institutions, and any company dealing with the public or the healthcare systems and food systems design corporations or the alternative health markets.

Suppose you weren't interested in any business but wanted to check out the reviews and comparisons of DNA testing companies at a Web site. So the possibilities are limited only by your creativity. If someone tells you that you haven't the credentials to do something, start a Web site reviewing and comparing the kind of companies in which you are interested, and let the consumers give you feedback as well as the professionals and experts in that field.

There are two voices to be heard—the consumer's and the expert from the most reputable companies in any field. That's what informational sites are about—selling facts to nontraditional markets, reviewing, and comparing. There's room for more than a few of these informational sites.

Certain people have different responses to nutrition than the majority of people. Twenty percent of people respond one way, twenty percent another way, and sixty percent still another. Do you respond to eating fruits by getting a bulging belly and too much insulin in your blood? Your DNA wants smarter food. Here's the big picture: Your DNA has cultural, biological, and nutritional core identities. Are you ready to look for the smile in your genes? Do mothers know best what foods their children will respond to according to the rules of nutritional genomics? Are the rules in place yet?

Your genes express their biological and cultural components in health or disease based upon the type of foods you eat. Your genes express their needs based on the type of exercise you do to stay fit as well as the way you perceive the stress in your environment. Your DNA is alive. It's conscious. It's a whole you in miniature. I'm a medical journalist interested in finding out my own response to nutrition. So I began by questioning scientists and other experts in genetics.

How about your response to what you eat? It's okay to want to eat smarter by learning to interpret the how-why-when-where-who-what-where of your entire genome and apply it to practical use in eating healing foods and choosing if needed, helpful nutraceuticals. The goal is to get consumers interested in looking at their own picture of health at the molecular level. It's a fantastic subject to learn. Question all authority and think for yourself. Eating for your genotype is here at last to free you from food cravings. Or is it available to all equally—to poor and rich alike? I say it is if you look deep enough into the research available to the public.

If you could get a printout of your entire DNA genome, what would it tell you about smart foods? How far would you go, how much would you pay to stay healthier and to prevent or delay chronic disease? What would you do to preserve your privacy and keep your employer, your HMO, or your primary care physician from looking at your entire genetic profile? The details are in the DNA.

CHAPTER 2

▼

INTELLIGENT NUTRITION OR SMART FOODS? WHO MAKES THE RULES IN NUTRITIONAL GENOMICS?

When a mystery and romance novelist turns to writing nonfiction books on DNA, from nutritional genomics to archaeogenetics, most scientists with doctorates do a double-take. What's encouraging is that some scientists in nutritional genomics and in other areas of genetics are speaking out to me, as you can see in the e-letter below, reprinted with permission:

————Original Message————
From: David Crawford
To: Anne Hart
Sent: Thursday, July 17, 2003 9:00 AM

Dear Anne:

Kudos to you for standing up to the profit warlords, politics and professional jealousy!
I am also a "full-time volunteer who writes for the love of science."

Albert Swietzer said "a person who is truly happy is the one who has diligently searched and sought out how to serve others". That is our founding truth at Interface Medical Research, Inc.
We work as consultants to the nutragenomic and nutraceutical community. From basic research, product design and development and clinical studies, we are committed to serve the scientific community and public.

We would be tickled to be listed as a service provider in you book. Thank you!

Please use my letter as you see fit.

David S. Crawford, PhD
Research Director
Interface Medical Research, Inc.
545 Farr Avenue
Wadsworth, OH 44281

Thank you, scientists who were willing to talk to me about how foods, genes, and healthcare are linked. It's as if there's a pyramid with your genes at the top and nutrition—foods and nutraceuticals—at one end of the triangle linked to your healthcare at the other angle. It's a triumvirate of intelligent nutrition tailored to your genes.

What I like about the science of bioinformatics is that the field is about managing databases of biological data. The future of bioinformatics helps to ensure that the growing body of information from molecular biology and genome research is placed in the public domain.

Information should be available equally to self-taught freelance writers as well as staff journalists and to the potential consumer of genome testing and research. Information should be accessible freely to all facets of the scientific community in ways that promote scientific progress.

If you need to do some fundraising to establish a DNA bank or support group project, look to celebrities, retired celebrities, media and film producers, and entertainers. Each year movie stars frequently headline Hollywood fundraisers and similar events that bring in money, sometimes in the millions, to donate to causes, most often for various campaign committees or health causes.

You might make some phone calls to see whether celebrities might want to get involved in planning events that would bring people together to raise funds for nutritional genomics-related projects, perhaps on healing foods and individual genomes. You might develop a slogan such as "Don't let your genomee become your enemy." Think in terms of foods for the majority in the midst of individual genetic differences.

What foods are best for the majority, 60 percent of the population who usually follow food guidelines recommended by healthcare professionals, dieticians and nutritionists based on general physical exams? How do you respond individually to certain foods and diets? Does your immune system go down when you fast or consume sugary foods such as fruit juice, when you exercise or travel? What foods contribute to your well-being?

When you research a study, find out who has done the study and whether it has held up to the critics or was found to be flawed by someone credible. If the study cannot be shown to be flawed, consider it as evidence to explore further. For example, on July 22, 2003, the Bee News Services of the Sacramento Bee, a daily newspaper published an article titled, "Fish Diet May Help Seniors, Study Says." The sub-title read, "Weekly helpings may cut Alzheimer's risk, doctors argue." The article listed Chicago as the origin of the news article.

What's missing in the article is any mention of what study the news piece referred to. The subject of the news reads: "Older people who eat fish at least once a week may cut their risk of Alzheimer's disease by more than half, a study suggests." The only problem is that the study is not named.

The news article proceeds to mention that the study adds to the evidence that what you eat might influence your risk of developing the chronic illness. What study? It refers to "a growing body of scientific evidence." I would have liked to see the name of the study so I could read the details in a scientific journal that usually publishes abstracts of and articles connected with studies. Where can I find this "body of evidence?"

I want to read the research studies linking various foods to cutting risk of various illnesses for myself. How about you? Is the recommended food good for my genetic expression and everyone else's or only for the twenty percent of the population who can eat most diets and still remain healthy into old age? What about the other twenty percent who can't eat certain foods without damage to their arteries and organs?

What the article does discuss is that the evidence adds to accumulated information on the subject of reducing risk of developing several chronic illnesses such as cancer, Alzheimer's, or heart disease, if people eat a diet rich in fish, fruits, and vegetables and low in saturated fats from red meat.

There are statistics from the Alzheimer's Association in the article. Approximately four million people in the USA are afflicted with Alzheimer's now, and Alzheimer's cases are expected to rise to 14 million by 2050. The evidence presented in the news article was that 'researchers' found "that people 65 and older who had fish once a week had a 60 percent lower risk of Azheimer's than those who never or rarely ate fish."

Who did the study? Who are the researchers? When was it done? No mention was given of the amount of mercury in the fish eaten. The article mentions that the meals included fish sticks and tuna sandwiches. No mention of whether the tuna was canned, canned with salt or no salt added, canned with oil or water packed, or whether it was fresh tuna or how it was prepared—grilled, fried, or baked, or boiled? Was the research done by the fish industry? What about the warning labels in supermarket's fish departments about the mercury in certain species of fish?

My neighborhood supermarket has a big sign posted at the fish counter for women of child-bearing age or pregnant women to avoid certain types of seafood with the names of the species mentioned due to the mercury levels. I see an article in the August 2003 issue of *Reader's Digest,* a cover story titled, "Hidden Dangers in Healthy Foods," that connects fish eating with individual reports of nervous system problems, illness, and hair loss due to mercury residues in fish.

Science writers should be accepted as media professionals by the science community without discrimination as to their credentials and should have equal access with all staff media of large newspapers, magazines, or broadcast networks to the growing body of information in genome research or any other area of molecular biology, especially areas related to preventive health and nutrition, foods, and vitamins and minerals.

The bioscience communicator whether self employed or staff employed acts as a go-between, a liaison between the consumer and the healthcare system or research laboratory, a type of educator and ombudsman. The communicator should be respected and allowed to access databases of biological data. Making the complex easier to understand for the consumer is news.

The scientific community should include journalists who specialize in writing about specific areas of science such as genomics without requiring writers to have science degrees. We attend enough genomics seminars, read enough journal articles, observe conventions, and read enough monographs and books to know what questions to ask the experts. In addition to looking toward the media as a liaison and ombudsman, the consumer of genetic testing is concerned about privacy issues.

For nutrition purposes only, results of tests for genetic risks should be private. The whole idea is to lessen the risk before a chronic disease develops. You can change how your genes express themselves based on food and nutraceuticals, but you can't swap your genes for some other genes.

Nutritional genomics is all about effects of intelligent nutrition—smart foods—on how my genes express themselves. Don't laugh at me. Smile. I need to know how specific dietary chemicals will nourish me at the molecular level. You also need to know about the effects of food on your own health.

Your genetic information—a printout of your entire genome—along with a profile interpreting it in terms of risk, disease, and recommendations for foods and/or supplements, should be kept private to be shared only by you and your healthcare professionals, and perhaps your heirs in a time capsule about DNA and medical history to be passed on to future generations. It's about what makes you tick at the molecular level, about DNA and health, behavior, and mood...If you are what you eat, why should you eat that way, and what will happen to your health at the molecular level when you stray?

Do you metabolize your meal fast or slow—burn your calories quick or over hours? Your genetic profile will give you clues, but you need to learn how to interpret it. And for the time being, you'll find it difficult to learn for yourself and take charge of your healthcare, because getting your entire genome tested is done in reputable places under the supervision of a physician managing your healthcare. You'll have to work hard to take control of your diet and your genes.

You have to work towards taking control of your genetic testing. The only way to do that for now is to learn how to interpret a genetic test of the entire genome and how to apply that to a diet that works for you. You need to get control of your genes. They are yours. The food you eat is yours. Why should putting diet and genes together cost a fortune now and be cheap in a decade? Knowledge is not only power. It saves you a bundle.

Okay, so the doctors need to eat, too. I should know. My son and son-in law are physicians. Just ask their wives about eating smart foods. The point is there are ways to learn how to look at a DNA test and look at a diet and understand how your genes respond to what you eat. Books are online and in the medical libraries. Read them, you autodidactic learners. Educate yourself. If a social worker can take a two-year masters degree program and become a genetics counselor, and if a nutritionist and a physician can learn to talk to people about what diets are healthiest for them, then you, too can learn to interpret a DNA test of your entire genome.

The trick is finding someone who comes highly recommended to give you your genomic testing. You do need to send your DNA to a lab. Where do you start? Talk to professionals in the field and find out what books and journals they read in order to understand how to interpret the DNA tests. Find out how your genes respond to what you eat. Very few people have the time to be interviewed, but keep asking around.

Find out whether anyone in the field or recently retired will teach an extended studies course for the public. Go to conventions and conferences where professionals in nutritional genomics congregate. Make friends. Read my romantic intrigue novel series on DNA. Or my books on tracing your ethnic DNA. Cook for your genes.

Have a nutritional genome feast/party. Invite people who can help you match your DNA to your diet or apply for a job in the field. Whatever you do, get a handle on, control of what you eat and how what you eat influences how your genes respond to that food.

Write about your dieting experiences. Go on the radio. Talk to people. Read professional publications. Sign up to receive the news of the industry. Ask yourself whether you want to get into an allied business from another angle where you'd rub elbows with people in the field. However you choose, get your entire genome tested, and find out how your genes respond to recommended diets compared to what you eat now. The whole industry is moving toward this becoming routine in about ten or more years. If you're as old as I, don't wait, start asking questions now, and find out what foods are healthiest for your genotype.

How does what you eat affect the expression of your genes? Nutritional genomics is causing a rhumba of rattlesnakes in the media. Do your research before you spend money on testing your DNA. If you don't know how to interpret your DNA test for nutrition, you'll need to find a physician who works with nutritional genomics professionals in the medical field and has had training in nutritional genomics.

For example, if the physician doesn't have a working contract with nutritional genomics professionals from reputable companies, how will the individual interpret a printout of your entire genome in relation to tailoring your foods? How many physicians today are trained in nutrition let alone nutritional genomics, a field that began around the year 2000?

So check out your healthcare professionals and the companies doing the research to make sure what they offer is what you want. Read the medical journals connected with nutritional genomics. You can find them in local university medical school libraries open to the public. Read some of the latest journals in the field. Go to conventions and attend meetings of associations of professionals who are connected with nutritional genomics and ask for reputable referrals. You can network with people in the field. Read about what research is being done by various companies.

When I was single and in my twenties, circa 1962, I joined clubs for single professionals where the scientists and technical people congregated so I could talk to them and enter the awe of their world in order to write about the aura of enthusiasm that engulfed them in such important work. I majored in professional writing/English and did get an M.A. degree, working my way through from the age of seventeen with a full-time day job, but since the age of eleven I had a deep interest in reading and daydreaming about science.

In 1962 my best female friend majored in chemistry and math, and I felt stifled by my day job typing important men's manuscripts and finishing my degree at night. Through chatting with the experts—the scientists, I found the inspiration to play with words and channel creative expression to novels featuring exciting, learned characters as scientists. I mean, they actually talked to me, a retired janitor's kid whose mom was a maid. (No—none of them proposed marriage, but I got to write romance and suspense novels about fictional scientists.) I needed something more—science news.

It was the high point of my life. Back in 1962 I ask each scientist who asked me to dance whether there was a new way to look at disease. I wanted to know then what my genes really wanted me to eat. What if genes could talk? What would they whisper to each of us? The DNA helix was big news with Watson and Crick visualizing the double helix. I ended up visualizing a double helix myself—writing the romance novel titled, *The BrideWore a Double Helix*. The outcome was I wasn't left on my own. We talked about foods, as in how a quarter of a fried potato eaten each day for a year could cause someone with "specific genes" or "fat cells" to gain 'x' number of pounds.

Our conversations on the ballroom floor were about consumers getting their proteins analyzed. More than forty years later, science has spectrometers, and you can find bioinformatics databases that manage nutrition-related information. You can now have a genetic profile compared and cross-referenced to your blood chemistry, metabolic tests, or any other areas of your physical exam or genetic profile.

You can look at genetic risks and find the foods and nutraceuticals that will help you. If you know eating fruit all day will make your blood sugar problems worse, because too much insulin is released into your bloodstream due to a genetic situation, you have some more control. That's what it's all about—control. It's you armed with ways to override your genes as much as you can with healing foods and nutraceuticals, if needed.

Don't only get tested and then left on your own on the dancefloor. You need some kind of medical supervision so you can learn how your test was interpreted and why those foods were prescribed as well as the effects those foods will have on the expression of your genes. You need to look at yourself at the molecular level. When you've self-educated yourself, you can take more control over what you eat knowing why the food affects you the way it does—at the cellular levels. You'll know why you're salt sensitive or not, and what happens inside your cells when you eat foods that make react the way you do.

Everyone takes nourishment expecting to feel well after eating. Don't let the caricature of yourself hold you back from exploring how you react to dietary chemicals in food. Look at the nuances, the changes in concentration of the nutrients you take in. Look at how the chemicals in foods make you feel. Eat certain foods and take your pulse and blood pressure. What makes your physiological responses healthier?

How long does it take between eating and seeing the results in your health? You need to compare your genotype, that genetic information such as your genome, your DNA, and look at the way you respond to a particular food or meal. That's why a lot of diet books don't consider your individuality at the molecular level. Many stop at the metabolic or chemical level or just look at blood types. That's important too, but so is your entire genome, all your genes and markers and DNA. What nourishment do your genes require for you to become healthier, delay the onset of chronic diseases, and feel well?

Your genes shuffle each generation. They recombine. Food, like DNA moves beyond family history, ancestry, and molecular genealogy. You need to realize the cultural, biological, and nutritional components of your genome in order to eat smarter for your genotype.

Here's how to eat bright for your genotype. The new field, blooming since about 2000 is called nutritional genomics. Nutritional genomics link what you eat to your physician-supervised DNA test of your entire genome and your healthcare. Nutritional genomics is about food plus supplements when needed and any medicine and/or supplements termed nutraceuticals.

Your genes will give out the signals of what whole foods are needed to delay or prevent infirmity. Who is going to test that DNA, and who will teach you to interpret the test when you can no longer afford a supervising physician for healthcare and have to feed yourself on your own on what money or lack of it may come?

Tailoring your food to your genome reminds me of what I heard for the past thirty years in the holistic food fanfare world of eating brighter for your individual metabolic self. Only today it goes beyond the metabolism and moves to your entire genome—your genotype. That's why a lot of diets don't work for everyone, because food affects people in different ways. What's one person's nourishment is another's allergy.

I've been touting eating healthy since 1959. It's what the holistic food and health conferences and conventions were spouting for at least three decades. Now it's arrived: eating bright according to your genotype. You have to eat smart, that is intelligent diets tailored just for you if you want to be healthy. That is, unless you're lucky enough to be part of the 20 percent of the population who can eat any diet and stay healthy longer.

Ever since I avidly read, tried, and was successful at eating for my blood type after reading about it a few years ago, and later for my personality type. More than a decade ago I tried to eat right and take the right supplements for my menopause.

When my genes acted differently to hormone therapy, I stopped the therapy back in the early nineties when my healthcare professional insisted that going off of estrogen and progestin was tantamount to a diabetic going off insulin.

Nevertheless, after looking at my physiological responses to progestin, I dumped the pills and took up exercise and raw veggies. What works well for another's genome, didn't work for mine. After reading a book on body type and diet, I cut out the sugar, being a "thyroid body type," and my large, overstimulated thyroid felt great again without the excess sugar. And since I began to eat whole grains and raw veggies, I tried to take charge of my own diet.

Before nutritional genomics, I had to read medical articles on blood type and diet or body type and diet, and figure out whether I was eating according to my genotype. Then I turned to books on eating for syndrome X, hyperinsulinism, and tried diets of 45% carbohydrates, 35-40% good oils, and 15% proteins to stop the excess insulin and insulin resistance after sixty from making my big tummy flatter on my thin, 124 pound frame. Without knowing my genotype, it was all based on metabolism, blood chemicals or blood type, ancestry, or body shape. Now with nutritional genomics, I can get to the molecular level, my entire genome…as soon as the price of testing becomes affordable to me…and available to consumers.

You can lobby for nutritional genomics to be available to everyone equally. How about the right to have your genome matched to a healthy diet? I want a good diet today that works for me. Why not match your genome to a diet that works to make you healthier? What's happening in this industry?

Consumer, be alert. I can't wait another ten years for nutritional genomics to be available to everyone on demand at little cost. I want it now. My solution is to write about it. My eight grandchildren will no doubt have it as their dads are physicians, and it will be a part of their children's lives. So what can you do now with your DNA as far as practical applications of DNA testing not only for ancestry or population genetics, but for nutritional genomics?

Scientists use the word 'interface' a lot as do business executives and people with day jobs. I have no day job. I'm a white-haired bag lady, so I don't have to interface. I interlace instead. What I do is greet you with a smile and hug. I don't interface with you. Only in nutritional genomics you have to study how to interface between your diet and your genes. Yes, that's you in there in the middle right between your diet and your genes. Now interface. Take a picture of your gene expression. Now smile. Snap. You've just developed a picture of your gene expression. It's profiled and digitally filed. You now have a picture of how you respond to what you eat. I'll say it again scientifically:

Nutritional genomics researches the interface between what you eat and your genetic processes. Scientists analyze your single nucleotide polymorphisms (SNPs)[1]. Your gene expression is profiled to get a picture of your response to your nutrition.

Individual nutrition is emphasized in the new field of nutritional genomics where you tailor what you eat, the nutritional supplements you take, and your entire health program of diet and exercise, work and lifestyles according to your own genetic profile—your genome. Forget about diets for large numbers of people.

Maybe those diets fit the sixty percent of us who can eat those things, but what about the twenty percent who need tailored menus according to our genes? Or what about the other twenty percent who can eat almost any food and still not develop the main diseases that could stem from eating the foods not right for our genotype—heart disease, hypertension, asthma, hardened arteries, cancer, loss of memory, and more? And are these degenerative diseases linked to changes in our genes? How do the different ethnic groups react to different foods?

All this research shines under the umbrella of nutritional genomics. This year it's tracing ancestry by looking at DNA test results, and next year the focus will

---

1. SNPs see http://www.ornl.gov/TechResources/Human_Genome/faq/
   snps.html. SNPs are "DNA sequence variations that occur when a single nucleotide (A,T,C,or G) in the genome sequence is altered." The letters are pronounced as "snips."

be on nutritional genomics, how to tailor what you eat according to your genetic markers. That's where the focus of my next book is. Most people aren't aware of the research being done at universities or at companies targeting research on nutrition and your DNA.

There are new fields within nutritional genomics such as the study of how people's DNA change according to their diets. There's the study of pharmacodiagnostic drugs, and phenomics, tailoring your medicine, therapy, and healthcare or exercise and lifestyle to your genes.

Books abound on eating right for your chemical, blood type, and metabolic systems. At the molecular level, you'll see books popping up on eating bright for your genotype, smart diets for your genetic markers. One diet book will not fit all. You have books on eating for Syndrome X, for diabetes, and other illnesses that appeal to groups, but what about eating bright for your genotype? Studies compare the immune systems of people from different ethnic groups or geographic areas.

This country faces a diabetes epidemic in children. Sugar is added to health foods such as soy milk or almond milk and other foods or beverages to bring people back by taste, even to 'addict' people to the sugar so they should buy more of the health food product in some cases.

Try finding an unsweetened almond, rice, or oat milk. You'll have to make your own from water and pureed grain. You'll have a hard time finding unsweetened soy milk, too but it's around if you look long enough at various shelves in health stores. Some health food products are drenched in salt so salt-sensitive seniors can't easily find a decent meat substitute that isn't loaded with salt, sugar, or fat. All this—in the midst of a revolution of research focusing on tailoring your food to your DNA to prevent or slow down premature degenerative diseases.

Make sure you're working with a reputable company that arranges to work with a managing physician. Beware of companies that test different aspects of your body to predict what to eat as the testing isn't of the complete genome and may not be done with a managing physician to interpret the results of the tests for prescribing your diet.

Don't waste money on incomplete or inaccurate testing. Watch out for testing companies that offer nutritional consulting based on genetic tests for a variety of complex diet-disease-or diet-health associations as they may not be reputable. Find out who is reputable before you spend any money. If you are testing for single gene defects such as PKU or hypolactasia, these can be tested for and diets arranged based on the test results.

Check out reputable companies that offer tests which are supervised. Some companies research the linkages between genes, diet and health. Don't go to a company that offers unsupervised tests. Learn the difference between a research company looking at how genes, diet, and health are linked, and a genetic testing service company. Beware of the unsupervised test or the company selling snake oil when it comes to finding out how your diet is linked to your genes and your health.

How do you know the difference? Talk to professionals in the research field before you think of any testing, and make sure everything done is supervised by a physician managing your personal healthcare who also is participating in research of how health, diet and genes are linked.

Think about all the Japanese who eat a dairy-filled American diet and land up with diseases not found in Japan when they kept the ethnic diet. What happens when other indigenous people take on a processed-food Western diet high in whatever disagrees with the nutritional expression of their genetic markers. It's a horizontal expression of a vertical desire. The remedy is tailoring your nutrition—foods and supplements—to your genotype as well as customizing your type of exercise and other lifestyle events.

Perhaps it's time to tailor our DNA to our multitude of ethnic diets and first find out which one we inherited or which diet works for our genetic expression. For me it's Omega 3 oils and low-carbohydrate vegetables, B complex vitamins, whole grains, and the Syndrome X diet, but how can we know for sure until our entire genome is passed through the test, and for now, the test is still expensive?

So we have to guess, perhaps at the mixture of ethnicities, family members and their lifelong genetic expression of what they ate. Look at what your families ate and what it did to them. Did you inherit any of those gene mutations? How has what you ate influenced your own health? The answer may lie in the SNPs, the genetic markers. Stay tuned for reporting on new research. These sciences include not only 'nutrigenomics,' but also tailored diets and understanding your genotype.

Related sciences that look at genetic variation to customize healthcare strategies include proteomics, metabolomics, bioinformatics, biocomputation, and phenomics. It's all about getting the details and the big picture of your nutritional status, requirements, and genotype. You can eat better according to your genotype without having to wait another ten or twenty years before testing your entire genome is affordable.

Start by observing your body's reaction to the nutritional and exercise environment. It's about personalized nutrition and medicine. Nutrients are not just nutrients. You have macronutrients, micronutrients, and antinutrients. The details come out in gene expression. Nutrients can alter your gene expression without changing your genes. Look at your metabolism, DNA, and gene expression the way you use your DNA test results to trace your ancestry. The door is open to personalize diets and customize your medical care. Under the umbrella of smarter nutrition, DNA plays a role that treats you as an individual rather than a member of a special group.

If you're an older individual, it's not possible to wait another decade to look for a customized smart diet to eat "bright" for your genotype. You can have your DNA tested now. Only be on the alert and research the company you're working with so you don't get scammed. A DNA test of your mtDNA or Y chromosome for deep ancestry is not enough of a test. Neither is a test of your racial percentages.

You need more clues—metabolic, chemical, and genetic clues. There are plenty of books on how to eat according to your metabolism or chemical clues, but you need more genetic information. What you need is your entire individual genetic makeup. If you can get a company to give your entire genome a pass-through and genetic printout, it could be helpful in preparing customized diets to help prevent and ease chronic disease.

The only obstacle is that you'd need either a professional to prepare the diet by interpreting your DNA test of your entire genome, and not all physicians can look at your genes and prescribe a diet. What you'd need would be a qualified, accredited, and experienced nutritional genomic specialist to consult with you, look at your genetic profile, and prescribe a diet. You'd need someone with enough experience in nutrition and genetics to know what foods would be best for you as a person. No diet fits all people, not even all ethnic groups. It must be individualized.

That's why it's called "intelligent nutrition." Smart menus are customized to your genomic profile. You eat at the molecular level. Eating smart for your genotype is science-driven nutrition. You can write a diet book for a specific individual, who's a member of a specific ethnic group of whom you can't prescribe a one-size-fits-all menu for that person within any group. There's too much individual diversity.

Where do you start? Research is the first step for any smart nutritional genomics consumer. You start by reading and going to conventions of food technologists and nutritional genomics professionals and/or students. Do your own homework. Start with reading about the metabolic diets based on an individual's chemistry. A decade ago, it was eating according to your blood type or metabolic type.

Now it's eating according to your entire genetic profile, your genome. The trend is becoming molecular, eating down to the atomic level in your molecules. That's because your genes express themselves at the molecular level, within and from the cells. The "eat according to your individual profile" movement began with alternative healing movements that always are keyed into scientific research at the genetic, molecular, and chemical/metabolic levels.

In most every alternative and holistic health magazine you find, the footnotes contain references to studies and medical journal articles. Check those out. It's a good use of time to learn how to read articles in medical journals and look up the terminology. If a study is flawed and is reviewed in another publication, read it. If a study holds up with time, keep a scrapbook of the study or articles for your reference.

Only now it's beyond looking at blood type or the lectins that agglutinate your blood from harsh reactions to foods, individual reactions. It's beyond chemical and metabolic, it's now genetic. You inherit and pass on recombination of genes that express themselves in various, individual ways, even within the same family. How you react to food is genetically determined and expressed not only in and through the genes, but also in behavior, mood, and sense of well-being.

Think of the potential for this science-driven food-for-health industry. You must explore the innovations and find professionals to network with to explore and evaluate the forthcoming innovations. Get on the mailing lists of the nutritional genomic research institutions. These research institutions may not do genetic testing of individuals for diets, but are engaged in the type of research you want to learn about before you put your health in the hands of a managing physician who must not only interpret your genome but prescribe diets and/or nutraceuticals.

You've heard of pharmaceuticals. Well, think nutrition and nutrition supplements. Research those nutraceuticals. Make sure everything is supervised, but that you still have an opportunity to learn about why and how your genes respond to what's prescribed, that is, you still have control over what goes into your body and knowledge of how it will affect your health. After you've finished being supervised for your own health, learn all you can about how genes and the diet work together to make you healthier. Then take control of your diet and celebrate your knowledge of a new language—a way to communicate with the expression coming from your genes.

Nutritional genomics is a science dedicated to researching smart diets customized for individuals with a purpose of creating a healthier population. Ask yourself as a consumer, how come it's so expensive right now to test the entire genome of a human to get a prescribed diet and so cheap to test the entire genome of a dog or race horse for breeding or diet? Lobby for costs to come down so diets can be prescribed according to one's genotype.

Besides the research scientists working or studying the field of nutritional genomics in academic or consumer consulting capacities, you have food industry leaders flocking to hear the scientists specializing in genetics and nutrition because the field is potentially a money-making enterprise for nutritional genomics-based food industries.

The whole idea of nutritional genomics got its wings around the year 2000 when studies revealed that a "dumb diet" as opposed to a "smart diet" (based on your gene expression) can fan the flames of your chronic disease risk. So to prevent or delay the chronic diseases for which your genes may be at risk, a diet prescribed only for your genes would help delay or prevent those chronic illnesses. You would need a printout of your genes based on DNA testing to find out for what diseases you may be at risk. Then the diet would be prescribed to prevent or delay those diseases.

Not only the diet, but the exercises and lifestyle and any other nutritional supplements would be recommended. The idea of an intelligent diet would be to nourish your genes at the molecular level based on what the genes needed to express themselves in the healthiest way possible for you as an individual.

So why wait a decade for science to come up with prescriptive diets? There's a lot you can do today with DNA testing and biotechnology before the impact on medical care, on the way foods are processed or not processed, and on your own health becomes influenced by the bottom line—profit.

Science has mapped the human genome. Back in 2001, you have the beginnings of the "marriage" between genomics—mapping your genes, and medicine. You have alternative medicine clawing at the door for decades demanding this knowledge and using it long before the medical fields ever thought of offering courses in nutrition beyond an introduction.

Now medicine and science is working to identify how genes work to change your health. In the meantime, the alternative health books have been emphasizing this all along—how to eat according to your body type, your blood type, your metabolic type, your chemistry, for years. Now medicine and the food industry is finally listening. The turning point came when the functions of various genes were identified with the aim of finding out how these functions affect your health, your risk of disease, and how the genes interplay with your lifestyle, environment, stress level, and even how you spend your day, let alone what you eat.

What you want to know now, especially if you're an older person who can't afford to eat the wrong foods for the next decade, is how and why your genes predispose you to sickness, premature aging, or obesity. While you're out there fighting for organic food, or worried about how genetically engineered crops will make you ill, here's one more concern: think how food is processed and how the process will express itself through your genes. Now customize your diet for your specific needs based on your genes—the expression of your genome based on the foods you eat and the lifestyle and exercise you do. In other words, eat "bright" for your genotype.

What is nutritional genomics? It's a market. It has legal and industrial implications for the food industry. We anthropology and bioscience communicators have been writing about alternative health for decades, exploring the whole oat grains and raw vegetables diets for years, looking at whether fish diets give us lower blood pressure or just a dose of mercury and writing about all sides of the whole foods spectrum, the details and the big picture.

Looking out toward the next decade, nutritional genomics professionals want to attract the baby boomers because of the large size of its population. However, the need is great right now among us parents of baby boomers, the senior silent generation. We are now over sixty and want to reverse our age-related conditions that pop up in the sixties and seventies decade of our lifestyles. We want nutritional genomics now, and we won't wait a decade. So how do we start, our little mtDNA or Y-chromosome DNA tests in hand that we sought to search for our deep maternal or paternal ancestry?

Don't you dare wait for boomers to demand smart diets prescribed for individuals according to their genome. We won't be scammed by diagnostics or counseling aimed at seniors by less than professional firms. We want the nutritional genomics professionals to be our entrepreneurs. To ensure this, we stick to the whole organic foods, but wonder which foods are best to nourish our genetic needs? Cooked, or raw? What specific foods are good for us as individuals, and who will prescribe them? How much will it cost, and how scientific is the information?

Could the genomics revolution actually be the alternative health movement and the whole foods movement joining together? Its purpose would be to look at the molecular expression of specific foods on the body. During the past 30 years, I've been attending holistic health conventions hearing lectures on the benefits of Omega 3 fatty acids, flax seed oils, fish oils, whole grains, low-glycemic diets for hyperinsulinism, syndrome X diets, raw vegetables, juicing with the pulp, the affects of too much fruit on triglycerides, how to flatten big bellies caused by eating sugar in people with hyperinsulinism, enzymes, food supplements, vitamins, and minerals, and what's needed to absorb calcium or magnesium, B-vitamins and TMG to reduce high homocysteine.

All of these are reactions expressed through specific genetic problems. And what I had to do 30 years ago was to take a specific, detailed look at how my body reacted physiologically, and change the food, customizing the food based on how that food influenced my physiology, that is the way my genes expressed themselves and reacted.

It always worked for me. Cut out the sugar, and the belly goes down. Exercise, cut the salt, and the symptoms change. Eat those whole grains that have a lower glycemic level, and the hyperinsulinism subsides until the next meal. Your genes are not my genes. What will work for you? Tailor eating to your own molecular level.

Know all about your genes and how food influences the way they express themselves. It is easy to guess at what the genes needed based on the symptoms and bodily response and also by looking at what the wrong diet did to family members who thrived on a Japanese, vegetarian, or Greek diet and succumbed on a high dairy, sugar, bread, coffee, canned tuna, hamburger, fries, too much yogurt, and chicken diet. The clue was to cut the calorie intake in half, and the disease retracted. So you got a handle on what the genes expressed.

A decade ago, I read books on the metabolic approach. I was a fast burner (metabolic), so I was supposed to look younger and thrived on eating protein and fat in the morning. Sugary cereals caused too much insulin and the shakes in the morning and weakness. Eating fish such as cooked salmon fried in olive oil mixed with egg whites in a patty with no salt, but celery seed, garlic, and onion powder felt healthy for breakfast. Oatmeal was too high glycemic and made me weak and shaky.

By working backwards from the symptoms, I could guess at my genes by what strengthened me after eating and gave me energy. Now I can work backgrounds from the general to the specific, from the effects of food on my feelings to the specific genes that put me at risk for the effects of too much insulin or insulin resistance and work backwards and forwards to prescribe myself low-glycemic foods that flatten my tummy and make me feel stronger and full of energy throughout the day.

I attended these holistic health fairs since 1970 when they arose out of the natural foods industry and the alternative medicine movement of naturopaths and nutritionists who were trying to get the medical industry to listen to them. Now they are listening as both sides—the alternative medicine and natural foods industries plug into the nutritional genomics research field from one end, and the medical and food industry from the other end. One group is visionary—looking toward the future. The other is benchmarking, based on what sold successfully in the past.

Prehistoric people ate berries in larger quantities than we eat today, when the berries were in season. Nutritional genomics researches diverse health effects for diverse genes. Flavinoid's effects on inflammation and on the nervous system are being researched. Ask yourself what can you eat that will protect your genes? Did prehistoric peoples have better nutrition then than the standard Western diets encouraged today? Did prehistoric peoples have similar genes to what we have today? Or did the slow-mutating regions of our DNA signal reactions to the change in diet and climate?

From the alternative food and health industries I learned about polyphenols. You find it in green tea, red wine extract, and other vegetable tannins. Anthocyanins are a type of polyphenol. According to Polyphenols Laboratories at: http://www.polyphenols.com/ main.php?PHPSESSID=976bfb699f4c396cfb032a25fa7877fd on the Web, "Anthocyanins is a large water-soluble pigment group found in a large number of fruits, vegetables and flowers.

These are the pigments which give plants their brilliant colors ranging from pink through scarlet, purple and blue. Some pharmaceutical effects of anthocyanins have been suggested, for example in treatment of cardiovascular diseases and in ophtamology. The antioxidant potentials of anthocyanins are high."

According to Marilyn Sterling, R.D.'s article in the December 2001 Issue of *Nutrition Science News,* "Eaten in large amounts by primitive humans, anthocyanins are antioxidant flavonoids that protect many body systems. They have some of the strongest physiological effects of any plant compounds, and they are also things of beauty: anthocyanins provide pigment for pansies, petunias, and plums." (Anthocyanins are a separate class of flavonoids from proanthocyanidins, discussed in *NSN* 2000;5(6):231-4.)

Back in folk medicine of the 12th century, bilberry, a bioflavinoid, was fed to young women to induce menstruation. Bilberry (*Vaccinium myrtillus)* is one of several anthocyanins. British pilots during World War II took Bilberry to improve their night vision. So if your genomic profile calls for anthocyanins, at least you'll know what plant pigments in fruit juices or certain wines or in nutraceuticals such as bilberry extract will be of help.

The genes of the plants, such as the genes in the plant pigments feed your genes. The work goes on at the molecular level. In the past, bilberry was used to treat ulcers. If it's found that bilberry may increase the production of stomach mucus to protect the bacteria that causes ulcers from attacking as bilberry was a traditional treatment for ulcers in the past, then folklore herbal medicines may have come full circle to play a role in genomic profiling.

See the Web site at http://www.medpalett.no/index.php?lang=en.

Then there are the procyanidins, found in grape seed extract and other plant extracts. I take grape seed extract in a capsule. I heard about polyphenols and procyanidins from the alternative health market and sometimes on radio shows that most people would call the "woo-woo" industry.

I heard all about grape seed capsules on the Art Bell radio talk show, the show that showcases UFOs along with holistic health speakers and topics of past life and so on. Sometimes you hear about these products in mainstream media and sometimes in alternative health media. The benefits of nutritional genomics has made it to mainstream mass media with articles in daily newspapers and the NY Times news magazine.

What you as consumers need to do is to read the guidelines of the National Research Council. They are the group that offer guidelines of recommended dietary allowances (RDA) based on the results research. So with advent of nutritional genomics, research may open doors for those guidelines to shift.

Now biotechnology has joined up with talking business applications. The polyphenols found in berries is being researched to relieve arthritis suffering. For the last three decades the holistic food conventions have been touting the same blueberries, grapes, raspberries, blackberries, cherries, to help relieve the pain of arthritis. Grape seed extract which contain procyanidins have been said to help strengthen capillaries.

Now science finds out that eating compounds such as polyphenols found in dark berries inhibits the expression of a particular gene. You still have the arthritis, but the pain lessens or goes away. Does it work? Try it and see how the expression of your particular gene reacts to polyphenols.

The procyanidins in my grape seed extract and the polyphenols in the berries or green tea I consume daily not only strengthened my capillaries, it got rid of my 'rhoids. Then there's body shapes. In Ayurvedic medicine of ancient India, body shapes are divided into vatta, thin, pitta, athletic/muscular, and kapha, rounded, with a layer of fat under the skin. Vatta is less prone to arthritis and more prone to anxiety. I'm a vatta.

So just by looking at body shapes you gain a handle on something about the expression of your genes. In the health food stores there are recommended food and teas for vatta, pitta, and kapha types. Just read Deepak Chopra's books, and you'll learn.

The wisdom of the ancients around the world knew something about genetic expression. The body shapes are genetic and react differently to foods. A vatta sipping coffee will sometimes feel anxiety from the caffeine, whereas a kapha may need the jolt to the nervous system to get going. Even without a DNA test, you can look back at reactions to food from the various body shapes and see what foods hit you like a nerve-shattering bomb of hyperinsulinism after you eat, and what food calms you or makes you more alert.

Before you try anything, read the studies and find out which, if any, are flawed, and which stand up to science. Does it work? Only your individual gene can tell you. It speaks not in words, but in the expression of pain or no pain or in other symptoms of strength and energy or weakness and fatigue or chronic illness symptoms. Will your physician work with you? Who will? What about training physicians in nutrigenetics consulting? Will other professionals fill the gap?

What nutritional genomics needs are companies that will sell food that is good for our body organs. Right now I walk into a health food store and find the healthy alternatives to the milk that makes me sick filled with sugar, rice sweeteners, fructose, corn syrup, or other additions, including salt that makes me sicker than drinking milk. Soy milk marked 'plain' is not plain. It's sweetened. You have to look for unsweetened, and most supermarkets don't carry it because they answer when you ask, "Will it sell if it doesn't taste sweet?"

Why are food companies addicting us to sugar knowing we'll come back and buy more of the product because of sugar cravings not because we want the healthier, alternative products. So you have to make your own to leave out the sugar, salt, barley malt, rice flour, or other additives that sell the product by addicting you to the taste rather than selling the healthy alternative point.

A few physicians sometimes also produce and sell vitamins, minerals, or vegetarian food products. Are these products best for your particular genome? Does one size fit the masses or only a percentage of people? My last resort was to make my own soy, oat, rice, and almond milk by cooking the grain in water and pureeing it in a blender. That way I can drink some almond milk by blanching the almonds and removing the bitter skin, then putting it in the blender with water. I soak soy beans overnight, puree them in a blender with water, then cook for 20 minutes and strain into a bottle. Soy milk has to be cooked. Foods have to be researched. Some types of sprouts may contain salmonella or other bacteria on the seeds.

Food companies need to join forces with nutritional genomics specialists who need to join with physicians, HMO-employed nutritionists, and other health care professionals as well as with genetics counselors. If the food industry doesn't promote food good for specific parts of the body, the people who produce food supplements and other health and diet tools will step in and plug into the genomics field.

If all would work together, food could be produced to normalize cholesterol. However, there is a need to create foods for those who need special foods such as low-salt, low-sugar, and low-fat foods that don't take out one and put in the other. For example, low-fat foods often contain a lot more sugar and/or salt than regular food products. Again, taste is being sold, not health. Look at soy cheese.

Most of it contains 290 mg. of salt. Regular Swiss cheese often contains only 40 mg of salt. Or look at labels of some fat-free cookies with lots more sugar and salt added than you'd find in those unhealthier cookies full of trans-fatty acids, true, but less salt and sugar. My answer is to bake my own cookies.

I take a cup of oat bran mix it with a half cup of flax seed meal and some cooked whole oat groats, add a few almonds, a handful of millet soaked in carrot juice overnight, and sweeten with fruit juice or a banana. I add two or three egg whites and a quarter cup of lecithin granules and a little soy milk if needed. Then I roll out the cookies and bake until light brown at 350 degrees F.

Before you boy low-fat organic alternative food products such as soy milk, look at the label and see whether the sweeteners added will send your blood sugar up. If you are genetically at risk for diabetes II in your mature years, try making your own alternative milk products from soy or almonds or grains such as oat or rice. Or look for unsweetened products when you can find them. Talk to supermarkets and convince them that unsweetened products are not marked 'plain' in many cases, and that they will sell, especially to the senior citizens who often don't know where to look for unsweetened or no salt added products.

Know your genes. The old adage "know thyself" means know how your genes express yourself. There are not enough specialized physicians to go around to all the baby boomers who will be knocking down doors in a decade for healthcare. Don't wait for the boomers. Senior citizens must take nutrition into their own hands and do the research on their genes now, not in a decade.

If you're retired, it's time to make nutritional genomics your main hobby as far as researching the effect of food and exercise on your genes. Nutritional genomics is linked to many other scientific areas of study. It's the most important tool for self identity ahead of personality tests and DNA tests for ancestry.

Ally yourself with others researching all the disciplines and branches of the life sciences. Food industry people need to read up on nutritional genomics. You don't need a special degree to know that food changes the way your genes express themselves. The prescribed appropriate and specific food combinations won't change your genes, but it will allow what genes you have to express themselves in the healthiest way they can. Some people may need to change their food or nutritional supplements based on their gene expression.

You could be taking vitamins and minerals in the wrong amounts or emphasizing the wrong supplements. Find out which ones you need and how much according to your genetic needs. Again, you don't have to wait another ten years until this is available to the masses. Do your research homework.

There are ways to find out what works with your genes. Join organizations that focus on allying people in the life sciences. Join the various life sciences alliances. People in many disciplines need to join together. For example the food industry, genomic scientists, healthcare professionals, journalists in bioscience communications, and the alternative health care markets need to plug into one another's research and resources. What's missing from these alliances? It's the consumer. Take charge of your search for self-identity at the genetic level.

If you don't see a new development and you're a consumer, perhaps you can develop it yourself. You don't need a degree in nutrition to make your own soy milk in a blender if you see the food products on the market are adding too much sugar and salt or barley flour to the milk you see on the shelves. Write to the companies. If nothing happens, make your own nutritious food.

The future of nutritional genomics is based on whether it becomes profitable. If it doesn't become profitable, no it won't be shelved. It will move to the holistic health markets and the alternative medical markets. The food industry is quick to move in with its own strategies, but it will always be profit as long as there is profit in healthy food. The mission is to convince the buyers in the supermarkets that people will buy food that isn't loaded with sugar, salt, and flour when the fat is taken out.

Nutritional genomics can plug into the phenomics industry—the science of customizing and tailoring your medicine to your genome. The U.S. diagnostics industry is a $2.3 billion business. You want your genotype? Your key-words in your consumer search for companies are "science-based." Contact science-based companies. Most do research, not individual genetic testing, unless it's for single-gene problems. So do your homework and don't bother the research companies if you want a genetic test. What you want to do is search for a managing physician who works with the reputable, scientific-based companies to look at your entire genome before any diet is prescribed.

You want to talk with people who are supervised under an umbrella and find out what research and testing is being done by whom, why, and how, where and when. Don't get taken in by snake oil-type unsupervised testing. Know the company and get recommendations from scientists and attorneys in the genome research business that work with supervising physicians concerning your healthcare.

You can look at a printout of your genes, your genome and conference with health professionals in consultation with you and your nutritional genomics consultants to find out whether a diet alone will help or whether a 'nutraceutical' supplement, medicine, or drug will help you along with the diet. The first place to convince and check out is your HMO and your insurance company. Find out whether you'll be reimbursed for the expense of having your DNA tested and for any sessions with genetic counselors. Again, don't wait until this becomes common and genetic counselors are flooded with boomers demanding tailored diets and/or nutraceuticals.

What will happen is either the price of the testing will come down or the number of people demanding it will make the price of health care premiums go up, or testing may or may not be covered by insurance in the future when the demand becomes high. You can help yourself now by forming alliances with genetic counselors and research scientists in nutritional genomics. Or at least read their research and join the message boards in nutritional genomics on the Internet. If you don't ask questions now, the answers later could become expensive or scarce, if the price goes down.

Make business alliances with people in nutritional genomics fields. What you're doing is taking research off the shelves of medical school libraries and bringing science to yourself—as a consumer. Again, you don't need a special degree to research this field for your own use. You can become a collector of published research in the field. Clip articles in the news or save Web-based articles for your scrapbook on this subject. Contact people in the news to answer your questions. Get your DNA tested—the whole genome. It's worth the thousand bucks or so now, even if the price goes down to a hundred later. If you get the right diet, ask yourself would it be worth it to your health?

What you want to find out about yourself are the links between your genes and your HMO or other health care systems. If you understand yourself, identify yourself at the genetic, molecular level, you'll think differently about your own health. You'll think differently about the health of your children and grandchildren if you think in terms of nutritional genomics as preventative medicine. It could delay the onset of what your genes predispose you to get. If you're at risk, instead of worrying, the dietary change could delay the onset or prevent it.

Time is on your side when you know how your genes express themselves. It's just the molecular you expressing yourself creatively. And since the body is hard-wired to try and heal itself, you could lend a helping hand with the specific diet tailored to your genes. A diet book can't fit a whole population any more than one shoe size will fit a whole group of people.

Think about your health at the molecular, cellular level, at the genetic level. It's all about using your DNA for ancestry in a new way, from genealogy to nutritional genomics. Now you can find out whether your vitamins and minerals or other supplements are actually being absorbed and whether you need most of them. If you do, then you'll know which ones are working and which are not.

Food industries, the agribusiness, genomics research, healthcare, and allied life sciences all are working together. The goal is to bring this to the consumer level now and not in the future. You have industries contributing millions to biotechnology centers. Many of these centers are university-based. It's time to bring the university ivory tower lab to the consumer. Food and drug industries are pouring millions into centers to further research. You have only a few scientists working on projects that bring the fields of genomics, nutrition, and health together.

You need more catalysts like this to bring people together to do more than 'interface.' You need consumer action to 'interact' with the scientists who are working in projects that 'interface' the genomics people with the nutrition researchers and the healthcare professionals and genetics counselors. You need consumer involvement and consumer-based research.

You need bioscience communicators to make the complex easy to understand by the consumer with little or no science background. You need the consumer to take charge of his or her own health, privacy, nutrition, and research. Extended studies courses would help bring the consumer into taking charge by learning what research is being done and comparing his own DNA genome test to various interpretations of the test by different professionals such as his or her physician, genetics counselor, nutritionist, and nutritional genomics scientist.

All these fields need to be brought together under an umbrella accessible to the consumer…not in the future, but now, and with the privacy issues intact for the consumer, and the research open to the public in publications or online to be read by consumers as well as by allied science professionals and the media.

Who will write the regulations? It should be up to the consumer to have privacy in individual genetic tests. Genetic profiles are the business of only the person taking the DNA test to be used for planning a diet or lifestyle change. And unless the tests are totally anonymous as to name and address, collections of genomes would benefit scientists studying ethnic groups, but only if there was total anonymity. Nobody wants to be labeled and have somebody else know the label can't be altered. It's too controlling. Privacy should come first. Privacy should be respected so people numbered not by name and address, but by anonymous numbers or letters would ensure the privacy of large groups of people.

Scientists would be eager to collect large numbers of genomes for databases. Your genes are private. They are your time capsule. Your genome test would be something you'd pass on in a family time capsule scrap book to your heirs or future generations to learn about what your genes expressed, what was in the family, and what may or may not have been inherited by heirs.

Each individual has a different genetic pattern due to recombination. Some genes are inherited, but not all. So a time capsule or book binder with your DNA test may be of interest for medical history to your grandchildren, but how much of this they have inherited wouldn't be known unless they had a test themselves and compared genes. The consumer could take an extension course or read about DNA and learn about what, how, and why, when and where the genes express themselves.

Check out the genetic testing products. Do your research so you won't get scammed by companies that could spring up anytime in the future anywhere and target certain customers such as senior citizens or Boomers. Learn which companies are reputable and check them out by satisfied customers and by the reputations of the people who work there. Make sure the companies have professionally trained and experienced people. Check with the universities and the genomic associations and centers to make sure you are being tested by a company they would use themselves.

Most people have no clue as to how to apply genetics to their personal life. Many fear privacy invasions. Few are told what risks and benefits surround eating bright according to your genotype. As regulations come into focus, ask yourself how to validate the testing. When you are able to validate the test, then you know you've learned enough to get tested. Always ask professionals how they validate the genetic test and learn from their answers. Then check out their answers with research and other opinions from various professionals in the field.

What can you teach yourself about the applications of nutritional genomics testing for your individual health concern? How do you apply the information you've read? How do you validate that information? If a company makes a health claim, find out how to prove it and test it before you pay for testing. Pass the word to parents and children about getting their children tested so genetic profiles will be used to prescribe better diets and exercise programs for children as well as parents. Go into the schools and speak at PTA groups on getting children tested.

With the epidemic of diabetes in children increasing, perhaps a genetic profile might help parents to plan healthier diets, beverages, and snacks for children. It might also help the fare in school cafeterias. The idea is to get the word out at the consumer level as to what research they can do as consumers regarding applications of testing and validation of the tests. Then what follows are the smart diets and diaries of changes in health for all ages. The time for tailored eating and lifestyle changes is now.

By the time we wait for cheaper DNA tests of our whole genome and the chance to apply the test results to a practical menu, in what shape will we be? Ask yourself, should what we eat need medical prescription? Learn for yourself. Powerful knowledge is out there for the public and the media. Take control of your genes, your food, and your nutraceuticals. Your destiny will be in the hands of nutritional genomic scientists until you learn the science of nutritional genomics and research how you are impacted at the molecular level by how and what you eat. Reactions to lifestyle, environment, relationships, and exercise, also are expressed through your genes and your body chemistry.

CHAPTER 3

▼

FOOD TECHNOLOGY, RESEARCH, AND NEW PRODUCTS OF INTEREST TO NUTRITIONAL GENOMICS ENTHUSIASTS

Research has to be applied to thrive, to interest business. Somebody has to fund genomics research for convergence to bring together and link genomics to food systems design and to healthcare. Will it be the largest international food technology companies that funds nutritional genomics as a research tool for better healthcare?

We go from the general—the body—the the specific—the genes, from whole foods to nutraceuticals such as concentrated extracts. You end up healthier and with your medical privacy, a special kind of virginity, intact. You also end up with whole foods tailored to your genes instead of tailored foods for the masses in a one-size fits all planet. Foods need to respect your diversity, including extracts and nutraceuticals. That's why exploring what your genes require for health is an innovation you need to learn.

According to Minneapolis-based Cargill's Web site at http://www.cargill.com/about/index.htm, "Cargill, Incorporated is an international marketer, processor and distributor of agricultural, food, financial and industrial products and services with 97,000 employees in 59 countries. The company provides distinctive customer solutions in supply chain management, food applications, and health and nutrition."

Cargill's media release of July 14th 2003 reports that food and pharmaceutical specialties are part of Cargill's Food System Design (FSD) platform, a group of specialty ingredient businesses that formulate integrated food solutions for customers. Cargill Health & Food Technologies, is a leading developer, processor and marketer of science-based, healthy ingredients for food and dietary supplements worldwide. H&FT is part of Cargill's Food System Design initiative in which Cargill businesses work with customers to produce ingredient solutions for affordable, nutritious, convenient and appetizing consumer products. H&FT is a business unit of Cargill, Incorporated.

FSD works closely with its customers to produce innovative products and services that deliver nutritious, convenient and good tasting food. It is a unit of Cargill, Incorporated. The idea of new product concept is the result of extensive consumer research. For example, Cargill announced a new product concept in its July 14th 2003 press release, a flavorful raspberry tea that supports bone health. It's one of the concept products unveiled by Cargill Health & Food Technologies (H&FT) at the Institute of Food Technologists' Annual Meeting and Food Expo in Chicago, that ran from July 13th–16th, 2003.

Suppose you had a DNA test of your entire genome for the purposes of nutritional genomics analysis and food prescription as part of a health program of diet, exercise, nutrition, lifestyles changes—the works. Assume you have just been counseled by your managing physician and nutritional genomics counselor that the way your genes expressed themselves required certain types of nutrients.

If a particular food or nutrient taken at a certain or any time is what would make your genes respond in a healthier way, delaying or preventing chronic disease now or later in life, then you might reach for what your individual genes might need. Perhaps it would be the refreshing prototype beverage, aptly named Bone Appetit. It contains AdvantaSoy™ isoflavones, Oliggo-Fiber™ inulin and calcium. Emerging science indicates that soy isoflavones may help maintain healthy bones and that inulin may help boost calcium absorption.

Let your genes speak to you by learning the language of your genes—your DNA. Learn or teach yourself what your genes are saying to you. They are alive. They speak. They are you encased in a molecular universe and make up every part of you. Listen for their message. What do your genes whisper in your ear? Is it a need for a prototype beverage? If so, what nutrients are in the beverage that you need?

Visualize your genes expressing themselves in the language of genomics showing you what they require. Up until now, you had to wait for your genes to communicate by your symptoms, by how you felt. Now listen to your genes by how they think. That's what intelligent nutrition is about, not only eating prescribed smart foods, but listening to your smart genes as well.

What nutritional genomics would do in your case is determine whether for you as an individual with a diet tailored to your genes would need a beverage containing soy isoflavones, inulin, and calcium. First, such a DNA test could help you determine whether your body would react healthier or not so healthy to those products. Let's say, it's just what your individual body needs. That's you.

Someone else may show different needs and different reactions to various products. The point is that the food industries have an open door to participating in the whole nutrition genomics revolution. It's like the Internet, with connecting links from your genes to the nutrients your body needs to operate at maximum efficiency. See the potential for a lot more convergence here of food industries, healthcare, and genomics stemming from cracking the human genome code a few years ago?

It's like the Internet making instant contact between your genes, your food, your healthcare systems, and the food industry. Convergence also works by going from macronutrition to micronutrition, from the wholistic you to you at your molecular level, your cellular level. By making right what's inside your cells, the outer you works better. Food system design and your genes are linked through a burgeoning industry and research arena called nutritional genomics. It starts with you having your DNA tested, not just for ancestry, but the whole genome, for links with the healthcare industry.

Now you have food system design, the healthcare industry, your genes, and the nutritional genomics research area linked, webbed, as compared to the development of the Internet. Convergence. You are now a catalyst, bringing various sciences and industries closer together. Business becomes science and science brings research in a more informed well of knowledge.

According to Cargill's July 14th press release, "The functional beverage segment is one of the most innovative in the industry," said Steve Snyder, Cargill H&FT director of sales & marketing. "Based on our consumer research, we know that successful products in this category will be those that taste great and meet consumer demand for products that positively impact health. Clearly, flavored teas fit well into today's healthy consumer's lifestyle. Our goal in creating Bone Appetit and debuting it during IFT is to spur exciting new product ideas with our partners in the food and beverage industry," he said.

Bone Appetit is made with AdvantaSoy™ Clear isoflavones. Proprietary processing technology results in beverages that retain their traditional flavor, color and consistency. "AdvantaSoy™ Clear isoflavones allow us to meet the challenge of creating new functional beverages that promote health and retain the delicious flavor and aroma which made them popular in the first place," said Snyder.

Cargill H&FT provides proprietary science and technology for food applications, new product development, product prototype development and regulatory support. The International Food Technology (IFT) convention attendees were invited to learn more about these products and partnering with Cargill Health & Food Technologies, when they visited booth #1748 at IFT.

Cargill's H&FT business is part of a larger Cargill initiative called Food System Design. As such, Cargill Health & Food Technologies works in concert with customers to produce ingredient breakthroughs and other food system solutions that result in appetizing, nutritious, and convenient consumer products.

Do you eat lecithin granules? I do. I need that particular nutrient for my health. What about you? The food industry also knows you sprinkle lecithin granules on your food for specific reasons related to your health. In Minneapolis, Minnesota and Stuttgart, Arkansas on July 9, 2003, Cargill and Riceland Foods, Inc., announced their intention to form a strategic alliance to manufacture, market and sell innovative lecithin products to food, pharmaceutical, and technical customers worldwide. Under terms of the agreement, lecithin produced by Riceland at its Arkansas (USA) facility will be marketed and distributed by Cargill. The agreement took effect by Sept. 1, 2003.

According to Cargill's July 9th 2003 press release, Riceland, a farmer-owned cooperative based in Stuttgart, Arkansas, has been a leading manufacturer of deoiled lecithin for 25 years. Riceland supplies high quality powdered and granulated deoiled lecithin products globally. Cargill Lecithin has a global sales organization and modern production facilities in Europe and South America.

"The Riceland alliance establishes Cargill as a lecithin supplier in North America," said Jens Heiser, president of Cargill's lecithin product line. "It also supports our goal of growing the lecithin business globally by serving our customers with premium products and the application properties they desire. Well-known brands like LECIGRAN™, LECIPRIME™ and LECIPERSE™ open up exciting development possibilities and are excellent additions to our existing business."

Cargill is a proven leader in lecithin marketing, logistics and identity preserved supply chain solutions. "This alliance is a major growth initiative for Cargill Lecithin and will foster innovation in the deoiled lecithin market," said R. Creager Simpson, president of Cargill's Food & Pharma Specialties North America, where the lecithin product line resides. "It supports Cargill's ability to provide breakthrough solutions with new, value-added products such as identity-preserved, non-genetically modified deoiled lecithin."

Riceland President and CEO Richard E. Bell said the alliance will allow Riceland to operate its lecithin manufacturing facilities at more efficient rates. "In recent years we've been handicapped by not having a source of conventional (non-GM) crude lecithin for processing. This new arrangement with Cargill helps us fill that gap."

According to the press release, lecithin is derived from soybean processing. In its refined liquid form it is used as a natural emulsifier. The deoiling process results in powdered and granular lecithin. Lecithin is used in a variety of foods as a blending agent, dough stabilizer, egg replacer and instantizing aid. It is also an important ingredient in pharmaceuticals, dietary supplements and cosmetics.

Riceland Foods, Inc., a farmer-owned cooperative headquartered in Stuttgart, Ark., has been a leading manufacturer and supplier of liquid and deoiled lecithin products for 25 years. It also operates soybean processing and vegetable oil refining plants and is the world's largest miller and marketer of rice. Its products are marketed globally under the Riceland and Chef-way labels, private labels, as ingredients and in bulk.

So you see the convergence here, the big, international food industries that design food technology systems are *listening* at food conventions to research done by the companies and universities engaged in research in nutritional genomics. Twenty years ago eating health foods was the butt of jokes. You've heard on mass media TV the clichés about how vegetarians and people who walk into health food stores look sickly. You've seen the look of disgust on some people's faces when invited to a vegetarian restaurant. Often they responded with another cliché, "Oh, you're not going to eat those twigs and sprouts?"

The tide has turned. No one makes fun of people who go into health food stores anymore. No one smirks at vegetarians who eat lunch at soup and salad bars instead of taking power lunches with three martinis, steak, fries, and a sedentary afternoon. Yet for some the soup and salad may not work well, and an hour later they are shaking with excessive insulin. On the other hand, the person eating the fatty steak and potatoes might have clogged arteries. So how you react to a meal is determined by your genes—your entire genome, with all its mutations.

Some people can eat any type of meal and stay healthy. It's all in your genes as if your genome is a deck of cards shuffling with each generation, except for your mtDNA and Y-chromosomes that pass on your remote ancestry key.

Even as more young people today demand natural food, those type of stores may also sell red meat because some people eat it. There's the competition in those types of stores. And with diets focusing on animal protein versus vegetable protein, the question remains: Will your genes express themselves in excessive insulin both with the high animal protein and high carbohydrate meals? And if so, what's the solution? The Syndrome X diet was helpful to me. What about you? The answer lies in going deeper, that is to the molecular level and seeing how your own genes react to different food combinations.

You'll hear warnings about avoiding the snake-oil companies popping up in the future or here already, but no one dared name them in print. You'll be told to research reputable companies. You'll find out about research companies and DNA testing companies, and learn the difference. Maybe you'll enroll in university programs of research or study and find internships or careers in nutritional genomics. The door is open, but where are the jobs? You don't always have to approach from a research stance into nutritional genomics careers. There are the patent attorneys.

The search for funding for research is real. Who has the money? In the meantime, know your genes. Are you ready for a preventive health profile not only at the chemical or metabolic level, but at the level of your entire genome? Don't forget your pet's genes, either. What food is best for your dog, cat, or race horse? Have you thought about what laboratories are researching in pet-related nutritional genomics? What genomic innovation is next? Let's take a look at ancestry and DNA.

If you're going to learn about how and what to eat for your genotype—your genetic signature—let's consider your ethnic origins. Food is a cultural component. And genes express themselves through cultural components. Genes have more than a biological signature. There's ancestry, ethnicity, and culture to consider when customizing foods. When the media mentions smart foods or intelligent foods, it's more than a perceived consciousness of your genes and the living food nutrients. What is really implied is tailoring or customizing food as you would configure software to your computer's operating system.

You have to tailor the food, choose the food based on what your genes require to function optimally. Since people are more or less diverse within their ethnic group, they may have inherited or not inherited certain genes that express themselves differently with certain foods. One example is milk/lactase tolerance or intolerance. If thirty-six percent of Southern Europeans can tolerate the lactase in milk and the rest cannot, that's one way ethnicity, genes, and food may interact.

If more Northern Europeans can tolerate the lactase in milk without digestive symptoms because ten thousand years ago a gene mutation in Northern Europe (Finland) allowed the people there and subsequently in most of Northern Europe to drink milk—tolerate the lactase in milk—it's a gene expression. If about sixty-eight percent of North American Ashkenazim and approximately sixty-six percent of Greeks can't tolerate the lactase in milk, it's a genetic signature.

How does your body handle certain foods? Find out what foods to eat from genetic testing and profiling. Tailored foods prescribed or recommended should be based on you as one individual rather than the ethnic group to which you belong. People of any one ethnic group are diverse enough to require individual testing. Let *your own genes* point the direction of your nutrition, exercise, and lifestyle plans, not merely the statistical averages of your ethnic group.

Your donated DNA may end up in statistical tables. Are you getting feedback for healthcare, or only a certificate printing out some of your markers? What practical applications can consumers find from DNA testing for nutritional genomics and also from testing DNA for ancestry? If you want to go beyond looking at DNA for deep ancestry or current DNA matches, what's out there for the consumer interested in human genomics?

▼

DNA Testing for Ancestry

Are you Curious about Your Ethnic Family Origins from Last Century Back 10,000 Years And How Their Genes Responded to What they Ate?

What your ancestors ate is part of your genetic history as a *cultural* component. How food they ate long-term affected their health is part of a *biological* component you may have inherited. People have different eating customs, choices, and habits. Lifestyle, environment, and exercise also plays roles. What did your ancestry and food history look like if you had to draw a map? How does your ancestry relate to the effects food, nutraceuticals, or medicine have on your genes?

Genealogists are now using molecular genealogy—comparing and matching people by matrilineal DNA lineages—mtDNA or patrilineal Y-chromosome ancestry and/or racial percentages tests. People interested in ancestry now look at genetic markers to trace the migrations of the human species. Here's how to trace your genealogy by DNA from your grandparents back 10,000 or more years. Where did they wander and camp?

Anyone can be interested in DNA for ancestry research, but of interest to Jews from Eastern Europe is to see how different populations from a mosaic of communities reached their current locations. From who are you descended? What markers will shed light on your deepest ancestry? You can study DNA for medical reasons or to discover the geographic travels and dwelling places of some of your ancestors.

What you're studying is non-randomness. You use DNA as a tool to study ancestry and the history of your ancestors as part of a larger population. You look for similar patterns.

Ashkenazim and Sephardim separated about 1,000-1,500 years ago. What happened genetically to each branch since that time? This book is about researching the possible origins of Eastern European Jewish genetic markers and DNA such as mtDNA and Y-chromosomes. It's a book that addresses the questions beginners have when studying how to interpret DNA test results for family history and ancestry. The focus is on Ashkenazic/Eastern European-derived Jewish ethnic groups. And there's much of interest to Sephardi and Mizrahi. The details here also are valuable for anyone else interested in ancestry research by looking at what DNA test results can and cannot reveal about origins and migrations.

You'll learn how to start interpreting your ethnic DNA test results for family history and ancestry. Who were the mothers, the female ancestors of the Ashkenazim and their communities in Eastern and Central Europe? From where did they originate or migrate?

How does this compare to the father's, the male's points of origin? What do the details in male and female DNA and genetic markers reveal about possible origins or migrations? Or is this impossible still to discern?

Scientists comment on a merging mosaic of communities that have since homogenized and are continuing to merge. Take the name Levien. It's an old Sephardic name listed from the book, *The Jews of Jamaica,* by Richard D. Barnett and Philip Wright. The book, *Jews of Jamaica* contains tombstone inscriptions and dates of death from 1663-1880. Only names that appeared as Sephardic are included in *Jews of Jamaica*. See the Web site at: http://www.sephardim.com/, a research tool for Sephardic Genealogy/Jewish Genealogy. There also are alternative spellings, including derivatives of Levy and the name, Lewin. There are also names such as Levit and Levita.

Also see: "Sangre Judia" ("Jewish Blood") by Pere Bonnin. A list of 3,500 Jewish names as created and defined by the Holy Office (la Santo Oficio) of Spain. The list comes from a census of Jewish communities of Spain by the Catholic Church. Names listed are those of Jewish origin.

Note that Levien, Levin, Levine and other spellings also are used by Ashkenazim and once were used and are still used by some Prussian non-Jews, the French using the spelling, LaVigne, as well as Spanish and Portuguese. Did the Ashkenazim take the name, Levine/Levin from the Sephardim or from the Prussian non-Jews? The name Levim also appears in the book, *History of the Jews in Aragon*, by Regne. Essentially a series of royal decrees by the House of Aragon, it contains Sephardic names recorded during the period 1213-1327. You're looking at 800-year old names that existed with the Sephardim and later—after the 18th century—also with the Ashkenazim in Eastern Europe. Levine existed in university records in Austria in 1598 and in Prussia in 1635 on Christian baptismal records, but is listed as a Sephardic name in the 13th century listings of Jewish names in Spain and Portugal. The pronunciation was more important than alternative spellings.

When and if you take a DNA test, you'll find that what's tested is your mtDNA or your Y-chromosome. It's affordable and easy to test that part of your non-recombining DNA for ancestry than the expensive route of trying to test your nuclear DNA that recombines. So only the non-recombining portions of your DNA are tested. Sometimes this part of your DNA is referred to as junk DNA. Note that your mtDNA or Y-chromosome only shows about 2% or 3% of your ancestry. The rest of your genes recombine and are much more difficult and expensive to test at this time.

What your mtDNA or Y-chromosome doesn't show is the rich tapestry of all your ancestors' origins. It just gives you the origin of your deep female ancestor when the mutations you have first originated. You're testing the matrilineal line—the mtDNA if you're female. For a male, he can test the mtDNA—his matrilineal line and also his patrilineal line. A male can test the DNA of his Y-chromosome to look at male ancestry possible origins. Women do not carry or pass on Y-chromosomes. What can a name tell you?

If you have a name that has a Prussian origin such as Levine, note that it also could have a Sephardic origin as Levien or other alternative spellings. Also, there were Sephardic migrations after the inquisition and sometimes before into Germanic-speaking lands, and the names could go back and forth. There is no such thing as a Sephardic gene, an Ashkenazic gene, or a Jewish gene. Jews from Poland have one of the most diverse genetic pool.

These genes or DNA existed before there was such a thing as organized religion or countries with boundaries. You almost always have a common ancestor who was another religion or ethnic group at one time or another in the past due to Paleolithic and Neolithic migrations and expansions.

If you're interested in an explanation of the DNA process with an emphasis on Ashkenazi genetic origins research, I highly recommend an excellent audio tape on DNA by Dr. Vivien Moses of University College, London, UK Microbiology Department, who with Dr. Neil Bradman, the geneticist, has focused on research with Ashkenazi DNA to see how diverse the Eastern European Jewish population is. The title of the audio tape is: <u>DNA and Our Ancestors: Are We All Jacob's Children</u> by Prof. Vivan Moses. It was recorded 9/8/1999 at 19th Annual Conference On Jewish Genealogy. You can purchase the tape at the Web site: <u>http://www.audiotapes.com</u>.

CHAPTER 5

▼

DNA TESTS FOR THE CONSUMER

With whom in the world do you share a common ancestry? With whom do you share a common ancestry and history? Looking at DNA to study ancestry or to study medical issues is all about tracing patterns—from patterns in the genes to patterns in the migrations. Where do the women—the founding mothers—of the Eastern European Jewish communities—the Ashkenazim—originate? How do their origins compare with the origins of the founding fathers, the patrilineal lines?

I asked genetic scientist, Dr. Mark Thomas, Department of Biology, University College London what his latest research findings were on Ashkenazi women as I had read his recent study.*The article referred to in his letter is titled, "*Founding Mothers of Jewish Communities: Geographically Separated Jewish Groups Were Independently Founded by Very Few Female Ancestors,*" American Journal of Human Genetics. 70:1411-1420, 2002. The study was researched by Dr. Mark G. Thomas, Martin Richards, Michael E. Weale, Antonio Toroni, David B. Goldstein, et al. Here is his reply.

Dear Anne:

Our study did not conclude that they are not related to one another and not related to anyone in the Middle East. Neither did it conclude that they are wholly Slavic.

Our study presented evidence for strong independent female founding events in most Jewish communities. This evidence was not as strong for Ashkenazi Jews and we proposed that Ashkenazi Jews were made up of a mosaic of different, independently founded communities that has since homogenised.

Because of the evidence for strong independent female founding events, assigning an origin to those founding mothers becomes a very difficult, if not impossible task. There was some suggestion in two communities, Indian Jews and Ethiopian Jews, of a local rather than a Middle Eastern origin for the founding mothers.

I have enclosed a pdf version of our paper.

I hope this helps.

Best wishes

Mark

—

Dr Mark Thomas
Department of Biology
University College London
Web: www.ucl.ac.uk/tcga/

Information I Received from DNA Tests

Some of the best genetic feedback material I received was from Roots for Real, a DNA testing firm that lets you see the geographic area of your DNA matches for ancestry from their Roots for Real mtRadius database. The map they sent me shows where in the world I have matches for my mtDNA, my matrilineal lineage. The idea is to see where my estimated maternal geographic ancestral roots would be on a map.

They enclosed an article by Dr. Peter Foster titled, "Mitochondrial DNA, the Peopling of the World and Your Own Haplogroup's Story." This article gave me some background in depth of how the world was populated. More important to me, it pointed out where my own specific genetic group or haplotype originated. The article is an excellent introduction for beginners trying to interpret the results of their DNA tests for ancestral roots study. For men, mtDNA or Y-chromosome can be tested for male and female ancestry. The fascinating marriage of genetics with archaeology provided an open door to view where my ancestors might have been 10,000 years ago—namely, in what today is the town of Bar sur Aube, France or the geographic center of my origins at 48.30N, 4.65E to the nearest location.

How could that be? I don't know anyone who is related to me that came from that area. I had been reared in a balmy, southern Meditteranean family eating the typical Med fanfare from grape leaves to feta cheese pizza, from olive oil to parsley and grain salads, and mostly vegetarian foods. My listening in music focused on Eastern Med quarter tones and Greek bouzouki. So home come my maternal line ends up mostly in Northern Europe? Fascinating. So what I discovered from that point, the tracing geography began as I mapped the route Mitochondrial Eve, the mother of all humans on Earth today took when she leave southeast Africa more than 130,000 years ago.

As a member of the leisure over 62-set with my time free all day, I have devoted/dedicated my working hours to tracing the geographic path of travel mitochondrial Eve took as I follow her descendants leaving genetic prints or mutations in their mtDNA as they moved from campsite to campsite from one end of the world to the other.

That's the awe about studying the peopling of the world. Because scientists now have genetic dating and DNA sequencing, you can trace your own ancestry from origin to present day migrations at least statistically. That's to Dr. Frederick Sanger, Nobel Prize winner, 1980, you can trace and understand your family tree by DNA mutations and genetic markers. It's a lot like looking at fingerprints people made as they touched the world to gain balance. It's also watching their footprints trot around the globe. In it, you'll find your own identity, or at least a few good possible places your ancestors might have been.

What you'll see are matches—people living today whose exact mtDNA sequences match your own, showing that somewhere back in the deep maternal ancestry you and they shared a single lineage connected from mother to daughter. The same can be said for looking at the male Y-chromosome about sharing a single lineage connected from father to son. Once you find your matches, you can see what geographic location they are at today and where the probable origin spot was geographically in the distant past.

After you find your matches, you are free to find out whether they want to be contacted or whether you want to start a group for people with the same matching mtDNA or Y-chromosome. Of course, privacy is usually asked for, but some people might want to meet their mtDNA or Y-chromosome matches. Somewhere in time, you might have shared an ancestor, maybe 10,000years ago or maybe 250 years ago or in-between. It's like looking at living archaeology joining genetics as a new field of archaeogenetics.

What science finds today may all be changed tomorrow. The field is so new, the new technology evolving so fast that data has to be updated frequently. You learn starting with genealogy and oral history and moving onto genetics that each haplogroup split at an early stage into small groups. The pioneers colonized the world and expanded demographically and geographically. In some tropical areas of the world, original DNA types didn't die out. If the population was large enough, and food was plentiful, the people thrived and retained an overall similarity to the original African ancestors. In Europe and Northern Asia, people were confronted with cold climate and Neanderthals.

Conditions were fierce, almost arctic with tundra and penguins in the Mediterranean. By 20,000 years ago only small groups could pioneer into Europe and Asia. Most took refuge in Southern Europe, in Spain and the Balkans. These areas became the homelands of haplogroup H and V. At the height of the Ice Age of 20,000 years ago, Native Americans were closed off from Asia by glaciers, leaving only a small group in Alaska until 11,400 years ago. Genetic variation continued to evolve in different areas of the world due to climatic changes. People followed the herds or the places where food and water were available as part of their migrations.

Take my haplogroup of mtDNA, H. It developed in a single female by 40,000 years ago who lived in the Middle East. That single woman had an ancestor whose haplogroup was N, and that N haplogroup had a female ancestor whose ancestor came out of Africa as L3 mtDNA haplogroup.

By the time the single female ancestor mutated to H, it was 25,000 years ago, and she first came to Europe in the middle of the last Ice Age called the Glacial Maximum. Her descendants landed in Spain and France by 20,000 years ago. By 15,000 years ago, her descendants colonized northern and Western Europe. So H is the most frequent Western European sequence currently and back then. Sixty percent of Western Europeans have mtDNA haplogroup H, and 38 percent of all Europeans, including the Eastern Europeans.

This is fascinating, but I needed to narrow it down to a place, and fortunately, Roots for Real sent me an actual geographic center with the latitude. They even mentioned the nearest location—Bar sur Aube, France. Now, that city is not scientifically assigned to my ancestor, but it's close enough for me as the geographic center.

I can see on the map by the coordinates that there was a migration north and south and slightly to the east, but mostly a straight line going from Spain and Portugal to Norway and Iceland and Scotland/England and from France to Austria/Germany/Poland/Bulgaria. Now if I could only figure out whether that ancestor is from Scotland/Norway, Spain/Portgual/Central Europe or Bulgaria? So I look at the geographic center for the most probable place of origin—Bar sur Aube, France.

Who has the most genetic variation of all the world's peoples? Africans who have ancestors living south of the Sahara. What's the pattern of variation in your ancestry? Variation decreases the further you travel from Africa—that is with people dwelling farther in distance from Africa. Long ago there was a bottleneck in Africa with only a few people surviving outside of Africa to migrate to the rest of the world over a long span of time.

So the more you have historic bottlenecks in an area which reduces the population to only a few, the less variation you'll see in descendants.

Different researchers such as geneticists and oral historians, archaeologists and anthropologists study very different areas of the human genome. They study different markers for a variety of reasons, focusing on different areas or loci. When different people study different markers for different reasons, all this information gets published in different journals, databases, and other sources of information. How do the archaeologists know what the geneticists are already researched which markers for what purpose? Yet science must compare populations to learn more about the history of how genetic markers traveled with migrations of people thousands of years ago and at the present time.

With all the findings scattered in a variety of publications that are not read by everyone, how do researchers learn from one another? The forensic scientists publish in forensic journals. The anthropologists have their own journals. The medical geneticists may be publishing and reading different publications than the oral historians, and the archaeologists may not be cross-referencing all those articles in one huge database. Take the Human Genome Diversity Project....It's great with coordination of multiple markers. You need to look at the whole world's population to find where you fit in. You need a database or data set. You can search populations at the data base called ALFRED. It's at the Kidd Lab Web site at: http://alfred.med.yale.edu/alfred/index.asp. For example, if you're European or Middle Eastern, you can do a general search or search at the Web site for a variety of ethnic populations, including the unplaced.

Where will your ancestors be found among which living populations today? You'll be surprised if you think you're ancestors are from one place and the matrilineal or patrilineal lines turn up in another area, showing deep ancestral origins before your ancestors migrated to where they were for the past few hundred years. If you're wondering whether your DNA found in Bulgaria means you are related to the Chuvash, it might be possible. The Chuvash language is the closest to the medieval Khazar language, and the Chuvash are descendants of the medieval Bulgars living in the Middle Volga region whose homeland was in the North Caucaus. See the publication:—Shnirelman VA. "*Who Gets the Past? Competition for Ancestors among Non-Russian Intellectuals in Russia*". Washington: The Woodrow Wilson Center Press (1996).

You never know, until your genetic markers give you a possible clue, but only a possibility, because your DNA may show up not only in Bulgaria but in northern France. Which is the real you?

Will my real ancestor stand up? (The whole world rises). I began looking for mtDNA matches from anywhere, especially Spain, Portugal, Greece, Crete, Bulgaria, Black Sea area, Italy, and anywhere in the Mediterranean. Most of my mtDNA shows up in Scandinavia, NE Europe, North Central Europe and Germany. So let's take a look at what the various DNA tests found.

I bet there were lots of mixing between Scandinavia and the Mediterranean. The Estonian unpublished database found my matches in Bashkortostan where the Bashkir population is perhaps often 65% Mongolian and the rest European. My mtDNA is also popular in Iceland. Take your pick. Iceland, Spain, or Baskhortostan? Our mtDNA H got around. Guess whose mtDNA most resembles the CRS in the entire world? No not the 20% you find in Europe.

The mtDNA haplogroup H with mutations is found at the rate of more than 45% in Europe, including the CRS)…but the CRS exactly is found 20% of the time all over Europe. Who has the highest match with CRS exactly and no mutations outside the CRS? Moroccan Sephardic Jewish people….27% CRS in one sample appearing in a research article.[1] In the same research article, Ashkenazim mtDNA matches the CRS in only 9.0% of individuals sampled.

Interestingly….my 356 mutation of H is found in central Italy and Armenia. With the other two mutations, 189, 362, is found everywhere else from Central Europe to Turkey….Iceland to Germany…and mostly…..ta, da…in Scotland and in the Orkney Islands at the highest rate. Anyone for a Scottish reel?

Here are my DNA matches by geographic coordinates. The geographic center is 48.30N 4.65E, Bar sur Aube, France with a deviation of 669.62 miles as done by "Roots for Real." Now that I have this information, what practical applications can I have with this information? For starters, I can make a time capsule for future generations.

Information on geographic matches of mtDNA for Anne Hart from the database of Roots for Real.

Individual Co-ordinates

Details

Roots for Real 2003 Page 1.

For further information contact:

http://www.rootsforreal.com

address: PO Box 43708,
        London W14 8WG UK

First, let's take a look at the mtDNA sequences.

My low resolution (HVR1) Haplogroup and mutations relative to the Cambridge Reference Sequence (CRS) are given below. A value of CRS indicates no mutations. High resolution (HVR2) results are shown. This test was done by Family Tree DNA. For further information, their Web site is at
http://www.familytreedna.com

HVR1 Haplogroup	H
HVR1 Mutations	16189C
	16356C
	16362C
	16519C

---

1. The article referred to in his letter is titled, "*Founding Mothers of Jewish Communities: Geographically Separated Jewish Groups Were Independently Founded by Very Few Female Ancestors,*" American Journal of Human Genetics. 70:1411-1420, 2002. The study was researched by Dr. Mark G. Thomas, Martin Richards, Michael E. Weale, Antonio Toroni, David B. Goldstein, et al.

HVR2 Mutations	263G
	309.1C
	315.1C
	523-
	524-

Note: The geographic The geographic center on the map Roots for Real sent me is 48.30N 4.65E, Bar sur Aube, France with a deviation of 669.62 miles as done by "Roots for Real." This is where the origin could have been anytime in the last 10,000 years of my matrilineal line. The geographic center is the nearest location where my mtDNa sequences show up. Best matches are on the map as red dots and the yellow center arrow on the map Roots for Real sent me represents the geographic center.

The exact sequences are in the Roots for Real Database (and other mtDNA databases) for my markers of mtDNA haplogroup H with markers at 16189C, 16356C, 16362C, 16519C. Note that the database is constantly being updated and only represents the sample. This represents only the motherline, the matrilineal lines. It does not show any racial, linguistic, or religious background, just the deep matrilineal ancestry. It represents people currently living in certain areas of Europe in the database with the exact mtDNA sequences as I have.

Individuals	Co-ordinates	Details
1	42.67N 23.30E	Bulgarian
1	51.50N 0.17W	White Caucasian, England & Wales
1	52.25N 21.00E	Pole living in Germany
1	52.00N 7.50E	Munster area, Germany
1	41.17N 8.63W	Portugal north of Douro, 20
1	38.72N 9.13W	central Portugal between Douro and Tejo, 21
1	47.27N 11.40E	Innsbruck, Austria
1	53.15N 18.00E	Bydgoszcz (Bromberg) region between Pomerania and Kujawy, Poland

1	53.15N 18.00E	Bydgoszcz (Bromberg) region between Pomerania and Kujawy, Poland
1	53.15N 18.00E	Bydgoszcz (Bromberg) region between Pomerania and Kujawy, Poland
1	47.98N 7.85E	Freiburg, S Germany
1	57.90N 5.17W	Scotland (NW coast) EMBL:AY024788
1	55.83N 3.07W	mainland Scotland EMBL:AY025191
1	55.83N 3.07W	mainland Scotland EMBL:AY024734
1	55.83N 3.07W	mainland Scotland EMBL:AY024537
1	59.92N 10.75E	Oslo, Norway EMBL:AY025926
1	59.92N 10.75E	Oslo, Norway EMBL:AY025925
1	59.92N 10.75E	Oslo, Norway EMBL:AY025924
1	51.50N 0.17W	England EMBL:AY025564
1	64.15N 21.85W	Iceland, accno AF236959
1	40.40N 3.68W	Spain

Here's my racial percentages genotype from AncestryByDNA: 97% European, 3% East Asian. (I'm supposed to take with a grain of salt anything 5% or less,) but maybe I am 3% East Asian? Anyway, here is my genotype printout. The numbers represent catalogue numbers, and the letters are my DNA markers called genotypes.

SAMPLE NAME = Anne Hart
SAMPLE ID = ANC0154
MARKERS GENOTYPES
    #958CC
    #960GC
    #961GC
    #963CC
    #964CC
    #966CC
    #969TT

#970CC
#971CC
#972CC
#973CC
#976CC
#977CC
#978TT
#979CC
#980TT
#993TT
#1000TC
#1015CC
#1022CC
#1029GA
#1033GA
#1036GA
#1041AA
#1043GA
#1044GG
#1047AA
#1048GG
#1049AA
#1050GA
#1051GG
#1053GA
#1055GA
#1056GA
#1057GA
#1058GG
#1060GA
#1062AA
#1064AA

This is the printout I received from Oxford Ancestors among other materials in August 2001:

**ATTCTAATTT AAACTATTCT CTGTTCTTTC ATGGGGAAGC
AGATTTGGGT ACCACCCAAG TATTGACTCA CCCATCAACA
ACCGCTATGT ATTTCGTACA TTACTGCCAG CCACCATGAA**

TATTGTACGG TACCATAAAT ACTTGACCAC CTGTAGTACA
TAAAAACCCA ATCCACATCA AAACCCCCCC CCCATGCTTA
CAAGCAAGTA CAGCAATCAA CCCTCAACTA TCACACATCA
ACTGCAACTC CAAAGCCACC CCTCACCCAC TAGGATACCA
ACAAACCTAC CCACCCTTAA CAGTACATAG TACATAAAGC
CATTTACCGT ACATAGCACA TTACAGTCAA ATCCCCTCTC
GCCCCCATGG ATGACCCCCC TCAGATAGGG GTCCCTTGAC

CHAPTER 6

▼

THE OLDEST NON-ARCHAIC HOMOSAPIEN mtDNA IN EUROPE

Did you ever wonder what diets the oldest Homo Sapiens ate based on DNA? The oldest mtDNA in Europe that's human, Homo Sapien and not Neanderthal or other archaic individual is U5. It had a common ancestor with its sister group, U6. The age of U5 is estimated at 50,000 but could be as old as 60,500 years. Where did U5 come from, as it's the first in Europe and evolved in Europe. The first place scientists find U5 in Europe is in Cyrenaica, and artifacts are found in Iberia. Syke's book says it shows up 45,000-50,000 years ago in Delphi, Greece. What effects did the fish and wild plants diet have?

The mtDNA U5 person had a common ancestor with the Berber U6, found in a third of Moroccans, which is its ultimate starting point before it arrived in the Middle East and then went on into Europe. U6 in N. Africa is close to U5 in Europe, and U6 is close in age to U5. The female who was the ancestor of U5 and U6 lived in what today is Morocco and Algeria.

U5 and U6 cluster with other Europeans and not with Sub-Saharan Africans. Today, U6 comprises a third of the Mozabite Berbers. There was gene flow between N. Africa and the Middle East. The ancestor of U5 and U6 lived in the Maghreb in N. Africa. U5 is found almost exclusively in Europe today.

U6 is found today in the Canary Islands, Iberia, N. Africa and Portugal. U5 is dominant in Scandinavia, particularly Finland, along with V and U4 there also. A large proportion of Canary Islander are U6.

The medieval Guanches of the Canary Islands also had U6. There was a lot of interbreeding in paleolithic times between U5 and U6. The Berbers are high in U6 mtDNA today. Whereas U5 today is found all over Europe and is the oldest European mtDNA, and is found more in Scandinavia, particularly Finland.

Not all Berbers are U6. The largest cluster of Berbers from N. Africa is H, especially in the city of Mzab. But when you test the Berbers with H, you find a sequence 16213 that has been found so far only in Europe, possibly suggesting a European origin for the H sequences in this N. African indigenous population of Berbers. They have numerous people with red hair and freckles, speaking Berber languages. The Kabyl of Algeria also have this trait. So did back migration to Africa take place in paleolithic times also?

Yes. Another common haplotype 16148-16343 belongs to the Berbers who have U3, a common haplogroup in the Middle East (Iraq) and also in Europe. And mtDNA U is found there, unrelated to U6, native to N. Africa. J is there, but J comes from Syria and Turkey. Some gene flow did come from further south in Africa, because a few Berbers are L3b, L2 and L3a, but sub-Saharan gene flow is only 14% among the Berbers of Morocco and the people of the Canary Islands in modern times.

So what this study shows is that European and Middle Easter sequences in the Berbers came from Europe within the last 10,000 years. Countries most likely— Sicily, Malta, and Spain. People from the Nile Valley migrated to Morocco in ancient times adding more mtDNA diversity. So what does it show? That U5 is the first echo out of Africa into Europe, but that it shows up as the first Europeans in two places, Delphi and Spain around 50,000 years ago.

For more information on this subject read, "The Emerging Tree of West Eurasian mtDNAs: A Synthesis of Control Region Sequences and RFLPs," American Journal of Human Genetics: 64:232-249, 1999, V. Macaulay, et al.

So U5 turns out to be the most ancient mtDNA in Europe (50,000 years to 60,500) and U6 in N. Africa. What's interesting is that U5 and U6 are "sister mtDNA groups" with a common ancestor in N. Africa. Each mtDNA group has a sister group.

For example H and V are sister groups, with a common ancestor. And J and T are sister groups. U and K are sister groups. Each sister group has a common ancestor that had in its signature both J and T or H and V or U and K.

Why the Deep Genetic Split Between Two European Groups of mtDNAs?

There's also a deep genetic split between some of the European groups. For example X is split deeply from H by certain transitions such as 16223T instead of 16223C in some, but not all X mtDNAs. There's a deep genetic split between (H, I, J, and K) and (T, U, V, W, and X).

What kind of event took place in Paleolithic times to cause this huge split between these European groups? Did the split take place before or after arrival in Europe? Was it the isolation of the Ice age that caused it?

Were the two groups separated, for example, in different parts of the world? It doesn't seem so, because H, I, J and K are in one group, and H lived in Paleolithic times in France and Spain, whereas "I" mtDNA haplogroup lived in the Middle East or Central Asia and J and K lived in Syria and also later, K lived in the Alps (from 17,000 years ago)...but also is found in the Middle East and all over Europe. So what split the two groups?

Look at the other group (T, U V, W and X). Note that X usually is grouped with I and W, and is found rarely in Europe and heavily in the Middle East and Caucasus, especially in Georgia. T is all over the British Isles, but also in the Arabian Peninsula and the Middle East.

U is found all over N. Africa, Europe, and the Middle East and is the oldest in Europe. and W and X, some in Europe, but most in the Middle East, N. India, and Caucasus, except for the X that went to the new World via the Central Asia through Siberia route, and is found among certain Native American tribes like the Ojibwa and Sioux, Lakota, and a few other tribes. What do you think caused the split between (T U V, W and X) and (H I J and K)?

Cro Magnons

If you like studies about Cro-Magnon fossils, read the article, "Morphological Evolution in Prehistoric Skeletal Remains," in the book, Archeogenetics, (Mc Donald Institute Monographs).

So, who were the Cro-Magnons? Their common ancestors were U6 from N. Africa and U5 from Europe. They had broad faces, were tall, slender, and long boned. Skeletal remains from caves in Spain such as Longar show they are most closely related to today's Swedes. The mtDNA studies on the Cro-Magnon fossils from the prehistoric Basques show they are slightly different from today's Basques, but today's Basques are similar to medieval Basques.

The Paleolithic samples showed they were closer to modern Swedes than to modern Basques. The Pico Ramos caves and other prehistoric Basque area samples showed the Paleolithic peoples were closer to one another than to anyone modern, but Basque population's ancient and modern did group together. mtDNA J was absent from the Longar site cave of Paleolithic Cro-Magnon samples, but the predominant prehistoric mtDNA was H, a high amount of H as if that's the dominant population there 22,000 years ago.

Other mtDNAs were identified—U, T and X. Interestingly, some other mtDNA haplogroups showed up that didn't fit in anything modern. Those just disappeared 22,000 years ago or so. Either they didn't survive to reproduce or they had only sons.

Fascinating....Even 20,000 years ago, H was still the dominant type in Europe as it is today—47% of Europeans are H. What was it about that group that had so many daughters survive to modern times, and what was it that made the other mtDNA groups smaller in size, at least in Europe?

H mtDNA haplogroup has been found to exist so far in only 6 percent of the Middle East today, but is the dominant type in the Caucasus at a smaller number than in Europe. Was it something in the food supply that didn't exist elsewhere in Europe at the time of the last maximum Ice Age? Or did more female infants survive? What happened so that today 47 percent or more of Europeans have H mtDNA haplogroup, including me? What has your research shown? Isn't reading about archaeogenetics fascinating and awesome?

CHAPTER 7

▼

FROM WHOM DO YOU DESCEND?

The Y-chromosome is paternally inherited, and it's a non-recombining portion that reveals gene genealogy of any ethnic group. Whether you're interested in human prehistory or your most recent common ancestor, that non-recombining portion of the Y-chromosome will in time show you accurate genetic genealogy. It's the large stable non-recombining portion of the genome. You have a large number of slowly and rapidly evolving markers. And your ancestry can be read in those markers from your prehistory to current yourself.

Wait, a minute. What about the maternally-inherited mtDNA haplogroup and its markers, mutations, substitutions, or transitions? Does anyone emphasize the important of learning about our founding mothers? Yes. When you see a table that shows a mutation at HVS-1 sequence site 355 shows up in a sample of 51.4% of the ancestors and present Georgian Jewish women, you try to visualize that single female founder with those particular sequences. She founded more than half the Georgian Jewish community. Keep in mind you're looking at a small sample. That's fascinating. I have a mutation at site 356. Would that be significant in looking down the tunnel to my own prehistory or yours?

The 355 mutation shows up on a table in that sample of only 1.0% of non-Jewish Georgians and 2.9% non-Jewish Syrians. In another table, my 356 mutation shows up among Armenians and Italians, (central Italy) but if I look at all of my mtDNA HVS-1 mutations, the entire sequence shows up in Spain, England, Austria, and Bulgaria, but also in Crete, Portugal, Scandinavia, Bashkortostan, Turkey, Germany, Croatia, among the Komi of Finland, Hungary, Poland, Albania, and along the Adriatic. In fact my exact mtDNA sequences are found from Iceland to Bashkortostan, with highest numbers in North Central Europe, North Eastern Europe, and in Scotland and the Orkney Islands.

Will the real ancestor of mine please stand up? I look at another database table from another part of the world and still more countries pop up showing me how many individuals in the sample of 10,000 people have my exact mtDNA sequences. Will time and research add more countries where my sequences are currently found in more tables from samples taken in different parts of Europe or elsewhere? Patience is a virtue when looking into the prehistory of your DNA.

If you look at non-Ashkenazi mtDNA, you'll find that studies by Dr. David Goldstein, Dr. Mark Thomas and Dr. Neil Bradman of University College in London et al. show that the women in nine Jewish communities studied show different genetic histories than the men. Their communities had a small number of founding mothers with little exchange after the community was founded with the host country. Their genetic signatures are not always related to one another or to those of the non-Jewish host populations of Middle Eastern countries.

That's not always the case with Ashkenazi women's mtDNA. If Jewish men had kept marrying local women over the centuries, the MtDNA diversity would increase among Jewish women. You have a bottleneck effect with the mtDNA of Ashkenazi women. When you look at the mtDNA Ashkenazi men inherited from their mothers you're seeing their mother's mtDNA. Yet 27% to 32% of Ashkenazi mtDNA is of the haplogroup K, and 9% of it is the same as the Cambridge Reference Sequence (CRS), which is haplogroup H, the most common haplogroup in Europe regardless of anyone's religion.

So where and when did the Ashkenazi mothers originate? Why do Ashkenazi women resemble their host populations slightly more than non-Ashkenazi women? How many of the non-Ashkenazi women were local women from the host country compared to Ashkenazi women? Scientists say that the early Ashkenazi mothers came from a mosaic of Jewish communities that homogenized and merged. Tradition says they fled oppressed lands to other lands that invited them. What do the genes say as scientists look at the details? Did traveling Jewish traders begin the early communities? From where did they seek their brides?

In the study by Dr. David Goldstein, Dr. Mark Thomas, you see in a table that 9.3% of Iranian Jewish women in that sample have mtDNA markers with mutations at 183C, 189, and 249. This shows up in 1.5 of Yemenite Jews but not in any Ashkenazim or in any non-Jews in that sample. In what direction does this type of information guide you? Perhaps you first look at the size of the sample.

For years the origin of Ashkenazim or those practicing the Ashkenazi rite and living in northern, north central, and north Eastern Europe had been a subject of many explanations. Some say male Radhanites, merchants left Baghdad and Persia to peddle wares, mint coins, or trade and travel, coming to Poland in the eighth and ninth centuries. Others say Jewish communities under oppression in the eleventh century moved East, formed communities, and intermarried with Jews arriving from Italy, France, the Rhineland, Spain, and the Byzantine Empire. Molecular remains (proteins and lipids) on stone tools and pots can tell us something about what our ancestors were hunting and eating.

The DNA tells us what groups of people were intermarrying with our family groups, and history may tell us why they came to join our settlements. Jews whose ancestors came from Poland are very diverse genetically.

People came from many countries to Poland in medieval times, and in the 18ᵗʰ century migrated from Poland to other lands such as Rumania, where they found Byzantine Jews from Constantinople living there, some refugees from the Spanish Inquisition of the 15ᵗʰ century. There also was a Jewish migration from Lithuania into the area around Bialystock, Poland around 1495. See the Web site on the history of Polish Jews at: http://members.core.com/~mikerose/history.html.

History tells us that in the third and fourth centuries, Jews fleeing oppression in the Middle East settled in Southern Italy from where they moved to Northern Italy and then into the Rhineland and then further east into what is the eastern part of Germany and then into Poland.

Still others claim that Jews living in the eastern parts of Germany intermarried with Sorbs, Wends, and Lusatians, all members of the westernmost branch of Slavs living in eastern Germany and Western Poland, developed the Yiddish language combining German words that replaced most of the Slavic ones, and finally intermarried with other Jewish communities moving into the same general area. The point is Ashkenazi DNA is a mosaic of the Jewish communities it has shared for hundreds of years, but where is its point of origin—West Asia, Europe, or the Middle East—or all three?

Studies of the male Y chromosome such as the one by DM Behar and K Skorecki, Technion, Haifa Israel researched 18 binary and 10 STR polymorphic genetic sites from eight Ashkenazi Jewish communities and compared them to non-Jewish populations from Europe in order to see what relative contributions of common ancestry, founder effects, admixture and genetic drift shaped the men's patterns and variations on the non-recombining (NRY) region of the Y-chromosome.

The analysis stronger supports a "striking homogeneity among all Ashkenazi communities with extensive inter-population male migration. The studies revealed "very low rates of admixture with non-Jewish European populations, and a founder effect for at least three out of the six major non-recombining (NRY) haplogroups."

I recommend reading the article, "Jewish and Middle Eastern non-Jewish Populations Share a Common Pool of Y-Chromosome Biallelic Haplotypes," Procedures of the National Academy of Sciences, USA, Vol. 97, Issue 12, 6769-6774, June 6, 2000. Then compare this article to those researching mtDNA and y-chromosomes for later years and see what new information comes in. Keep updated on the latest research.

New research seems to tie in findings with practical applications from molecular genetics to phonemics to looking at the entire genome to how this information may be used by oral historians and genealogists, by physicians, linguists, and by researchers in many fields. See the recommended bibliography at the back of this book and in other genetics or genealogy publications you find in libraries. Also you might find of interest linguistics professor, Dr. Paul Wexler's book, *Two-tiered relexification in Yiddish: Jews, Sorbs, Khazars and the Kiev-Polessian dialect*, Berlin-NY 2002: Mouton de Gruyter Press, or the book by Jeff Malka, M.D., F.A.C.S., F.A.A.O.S, author *of Sephardic Genealogy: Discovering your Sephardic Ancestors and their World*, Avotaynu, 2002. Dr. Malka was recipient of the *Reference Book of the Year Award* for 2002 by the Association of Jewish Libraries. Avotaynu, a publisher of books and a journal of Jewish genealogy has a Web site at: http://www.avotaynu.com/

What about the women? What do their genes show as to origin? The key is extensive inter-population male migration. Ashkenazim spent two thousand years moving from the Middle East to Eastern Europe. Along the way, the Jewish communities that lived in Eastern Europe until recent times had varying degrees of admixture and gene flow with their neighboring populations in which countries they dwelled.

Mostly, the admixture was between mobile Jewish communities. As the Jewish population left the Levant after 70 CE, they set up villages at intermediate locations along the way from the Mediterranean to North Eastern Europe. After traveling through the Levant, Anatolia, and Greece, they sailed to Italy. At that time (70 CE) Rome had a large Jewish population and some Romans and Greeks converted to Judaism, especially those living among Jewish communities or living in Roman colonies and villas in Palestine during the time of Roman rule.

Greeks in Alexandria may have converted to Judaism if they had close contact with the large Jewish community there in classical times. As the centuries past and Rome fell, the population of Rome decreased from a large, highly populated city with a large Jewish component to merely 25,000 people during the early dark ages after the fall of Rome when its Empire moved east to Byzantium. Some Jews moved to Anatolia to join the bustling cities of Byzantine Jews there. Some Jews kept moving north in Italy, although the Southern Italian peninsula provided a first destination for communities of Jews.

By the fourth century CE, Jews that were to become the Ashkenazi population moved north, settling in Northern Italy and living there until around the 10th century, when they formed communities farther north in the Rhineland. By the 10th century, Jews set up communities in the Rhineland (Germany) around the Rhine valley. This became the cultural center of the northern European Jews, the Ashkenazim as they kept moving north and east, expanding after the 11th century. As they settled in Eastern Germany and Poland, this shifting community gave rise to the largest Jewish Diaspora community to live in Europe before World War II.

Scientists have studied the heterozygote advantage versus the theory of genetic drift. Healthcare professionals have researched the recessive disease genes in the Jewish communities. Paternal lineages have been studied. What about the maternal lines? Where did they come from? Science looks at mutation rates and most of all, patterns. Research looks at genetic markers. Along come family historians to look at oral history or family traditions of who's a Cohen or a Levite and other information. Who contributed what genes to shape patterns?

Researchers study "polymorphic sites" to look at genetic contributions. The goal is to find common ancestry or study variability—diversity or lack of it and to look for founder effects, admixture, or genetic drift. So we know there is homogeneity among all Ashkenazi communities. We know the males traveled within the various Jewish populations, and we know the rate of admixture with non-Jewish populations is low. How much is admixture, and how much is genetic drift?

What about the women? Did the males take their women with them from the Middle East or marry local women in Italy, France, Germany, Poland, or Russia? Or did they travel the length of the Jewish communities from Spain and N. Africa to the northern most reaches of Germany, Poland, or Russia to find a "nice Jewish girl"? Did they marry local Non-Jewish women, mesmerized by the tall

blondes and redheads of the German forests or marry "Viking" or
Finno-Ugric-speaking women of medieval Kiev or of the Ural mountains? Who
were the brides of the earliest Jewish communities in Northern Europe?

It has been said that American Jews constitute about 27 to 30 percent of any
group whose distinguishing characteristic is intelligence. Scientists reveal that the
mean Jewish IQ is about one standard deviation above that of other Europeans.
Does this show anything genetically? You can play with unrelated statistics—
about 27 percent to 30 percent of Ashkenazim have mtDNA—matrilineal lineages with the K haplogroup. But does this mean anything other than random
statistics?

What you have to do is look at reality checks from what the genes actually say.
You can do tissue typing, look at white blood cells—leukocytes and look at ethnic origins. It's done by anthropologists trying to find out what ethnic groups
from which geographic locations match—for tissue typing reasons to help people
get the correct transplant or marrow match.

In the case of mtDNA looked at across three thousand years, there's a lot to
research. Too often when the word "Ashkenazi" is mentioned, the question is
asked whether they have absorbed the Khazars that converted to Judaism in early
medieval times and then disappeared shortly after their country was conquered
by the Kievan princes. Did they flee to Poland, the Ukraine, and Romania? Or
did they disperse and merge with the mosaic of Jewish communities across
Europe?

What happened to the medieval Jewish community of Armenia? Who else
contributed to the Ashkenazi population? Who converted to Judaism in the
regions near the medieval Jewish settlements? The story begins as Jews are pushed
out of Western Europe and flee to the East as oppression and the bubonic plague
decimates Europe. Few migration routes of Jews lead from East (Khazaria) to the
West. Most routes run West to East and from Lithuania to Poland or from the
Crimea to the northern Russia, from Byzantium to the Ukraine.

Immigration to Poland/Lithuania before 1600 was largely coming from Germany and Central Europe. Migrations from Hungary went east towards Russia.
Polish princes invited Jewish and non-Jewish settlers from Germany right after
the Mongol invasion decimated the Poles followed by the bubonic plague of the
mid-1300s. The situation in Western Europe was so bad for Jews in the 1340s
and 1350s, the years of the bubonic plague and Mongol invasions in the East that
Poland known in Hebrew as Polin, (Poylin in Yiddish) land of rest, looked like
an open invitation to great farmland and forests. The Ashkenazim moved east.

The Khazars didn't as one huge community of 500,000 people suddenly move West into Poland in the 10th to 12th centuries at the rate that the German, French, Central European, and Italian Jews moved East to escape oppression in Germany and/or N. France and the bubonic plague. Geneticists are not finding traces of Khazari genetics unless they are testing for medieval Khazari genes.

What shows up in some Ashkenazi males—possibly—are genes found in some Slavic peoples, such as the Eu 19 chromosome that appears in 56 percent of Polish men and 60 percent of Hungarians. But Eu 19 shows up in 10 percent of Syrians. The marker comes from the Gravettian wave of Paleolithic hunters who came to Eastern Europe from Western Asia and Central Asia 20,000-40,000 years ago. Western European males have Eu 18 in larger amounts. Eu 19 is a marker for Eastern European males. However, what also shows up in some Ashkenazi males also is the CMH, the Cohen Modal Haplotype. Some men carry these markers and some don't in the Ashkenazi population whose ancestors came from Eastern Europe. You have to look at where they were before they arrived in Eastern Europe.

What Eastern Europe was to Jews in medieval times was a refugium from oppression they found in Western Europe from Spain to England and especially during the crusades, the plagues, and before that, the fall of Rome that had a huge Jewish population. Southern Italy was the first route out of the Middle East into Europe, then northern Italy and onto the Rhineland valleys and France. When life in Western Europe became unbearable for Jews, they looked for lands that invited them. The princes of Poland invited them.

Poland has the most diverse Jewish genetic patterns in Europe. You find Sephardic Jews coming into Poland in small numbers, families with Spanish names such as Rangel and Angel. Rappaportes from Portugal and N. Italy. You have people coming from Lithuania and from Germany. Jews in Bialystock are diverse. Some still have ancestral features very similar to Assyrians or other peoples from N. Iraq. Only a look at the DNA will show to whom they are related the closest. You have to look at the genes before you accept at face value reports online that claim that Polish and Russian Jews are descended from Khazars.

Keep in mind that Ashkenazim are a mosaic of many Jewish communities migrating and merging, coming together to form a pattern. You have people from Persia and Spain in Poland intermarrying with people from Prague or Germany, France or the Ukraine, Romania or the Byzantine Empire. Marriage usually took place within Jewish communities, and travel was wide, especially by males from one Jewish community to another, especially for trade. Polish Jews traveled to Odessa in the Ukraine to find spouses, and Jews from Romania married Jews from N. Poland. So diversity exists, even as the mtDNA haplogroups go through a bottleneck.

Today you're told to marry outside your own Diaspora to keep genes mixing and the diversity growing. What about languages spoken by Ashkenazim? Yiddish was spoken in Eastern Europe, Ladino after the 15th century and/or Greek before, in the Byzantine Empire. Romaniots in Greece still speak Greek. These Jews have been in Greece since antiquity, later joined by Jewish refugees from Spain. Polish Jews migrated to Rumania in the 18th century and found Jews living there who came to Rumania from Constantinople in the 15th century. In medieval times, Jewish communities were founded quickly. Did the women come with the men and families? Or did the young men set out alone and find a local bride?

What will tell the story is genetics. You keep looking at the women's mtDNA and comparing it to Y-chromosomes of men until you see patterns that shape the migrations that tell a kind of oral history written in the genes. So even though most women's maiden names were not recorded (not many had last names until the 18th century), their genes identify them. They existed and have a core identity that lives on in their descendants. They were faithfully 'frum.' By 1600 nearly 20,000 to 30,000 Jews lived in 60 European communities of Eastern Europe.

Sometimes looking at the history of a language can give a clue to how the genes landed in a particular container. That answers the long pondered question of how Jewish genes landed in the most northern and eastern parts of Europe and on other continents. Yiddish has had an ancestor language it came from before it was Yiddish. So has Hebrew. You have Slavic, German and Hebrew words in Yiddish. You have a few ancient Egyptian words in Hebrew, such as Moses (from the water) in Egyptian.

What words does Yiddish have that are Khazari other than "davenin" for praying and perhaps yarmulke for kippot or skull cap? Linguists suggest some Khazar names such as alp for brave, but did the name Alpert or Halpern come from the Khazari Alp for brave or from the Germanic Heilbrun or Halpern or the Slavic Halperin? Did Kaplan, the common Ashkenazi name, come from Kaplan, a Khazari word for tiger or from another language? The people speaking the language most closely related to the Turkic dialect of the Khazars today are Christians living in Chuvashia.

You be the judge and do your research at the linguistic and Khazari history sites on the Internet. Then check out secondary and primary sources for historical research. Were the Khazari Jewish names originally Slavic? Is it important whether one or two people converted to Judaism from a Khazar, Turkic ancestry or from a Sorb/Wend/Lusatian/Slavic Polish-German ancestry? Or from a Middle Eastern ancestry? After all, 20 to 26 percent of the population of non-Jewish Europe came to introduce agriculture in Neolithic times from the Levant/Anatolia/Syria about 8,500 years ago and carried the J mtNA haplogroup. When they arrived, they met the Paleolithic European hunters. So 20-26 percent of Europeans have the similar Middle Eastern genes to what the arriving Jews are supposed to have.

It's true that Khazars living where the Volga meets the Caspian had royal families who converted to Judaism. However, after the destruction of Khazaria, the people living under the Khazars' rule were Slavs and Armenians, and for a time, Georgians. If these Slavic people also converted, they would tend to marry into the communities of Byzantine Jews living in their lands as well as in later centuries with the Jews moving east from Germany and Central Europe.

Whatever the DNA would show would be a mosaic, a rich tapestry of mixtures of Jews of Iraq, Persia, medieval Armenia, Germany, France, Italy, Greece, Turkey, and the rest of Eastern and Central Europe. Yet even the Sorbs of Eastern Germany and Poland have had an origin in Western Asia as well as Europe. So what can the DNA tell other than that after the communities of Jews were formed from the mosaic, they remained relatively isolated and intermarried within their group?

A bottleneck of mtDNA, of the female line resulted with few women, few founding mothers in the Ashkenazi communities. Today, you have the Himyar communities of Yemen, a Jewish group marrying with Polish Jews, Moroccan Jews marrying Lithuanian, and so on, especially in Israel.

In the past, dozens of studies already have shown the common genetic origin of Jewish communities by studies of molecular markers, and studies also have shown the admixture between Jewish communities and their neighbors (Mourant et al 1978; Livshits et al. 1991. Bonné-Temir et al. 1992). You can read these scientific articles and decide for yourself about the high level of paternally inherited Y chromosomes that reveal Ashkenazi and Sephardic Jews all have high frequencies of haplogroup J as pertaining to the Y chromosome, the male or patrilineal ancestry, and that both Ashkenazi and Sephardic males are similar in their Y chromosome haplotype J with Lebanese non-Jews. The populations of Central Europe don't have this pattern.

The Cohen modal haplotype (CMH) is a microsatellite haplotype within haplotype J in the Y chromosome that links Jews—both Ashkenazi and Sephardic to a common ancestor with other non-Jewish men from the Middle East. The CMH is a type of genetic signature of the ancient Jewish priests, perhaps of the ancient Hebrew people and also of many other people in the Middle East, Southern Italy, Armenia, Greece, Hungary, Sardinia, and other areas who had ancient contact with peoples of the Eastern Mediterranean.

Yet not all Jews have the CMH. For the ones that have the CMH, it doesn't matter whether they lived in Northern or Eastern Europe or in Spain or in the Middle East. On the other hand, in the early medieval past, conversions were common. The point is the communities mixed, intermarried, and took their religion seriously regardless of their common language or difference in language. They all had Hebrew in common as their written and sacred language.

What the studies didn't do is to decide which geographical origins brought forth the founding female mothers—the founding mtDNAs within the various Jewish groups. What studies did find out that the mtDNA of Ashkenazi women went through a bottleneck. And with the non-Ashkenazi women, the founder effects were so severe that they show a great difference from non-Jewish populations. This means there were very few female founders—at least eight—in the non-Ashkenazi population. However, in the Ashkenazi populations there were more female founders, still less than in the non-Jewish populations.

The non-CRS modal haplotypes in Jewish populations are rare in non-Jewish populations when you look at the women, the founding mothers. You can't assume that there were few female founders in the Ashkenazi communities, although there definitely was an mtDNA bottleneck in Ashkenazi shtetls or villages.

When you look at the genes, the DNA of Jewish women of Eastern Europe and Northern Europe who say they belong to ancestors who dwelled in Ashkenazi communities, spoke Yiddish, and practiced the Ashkenazi rite, you find that the Ashkenazi population has a modal haplotype with a frequency similar to the non-Jewish area in which it dwelled—that is the host population. The study found 9.0% compared to 6.9%.

This revealed that there wasn't a strong founder event for the Ashkenazi women. So if there wasn't a strong founder event for Ashkenazi Jewish mothers, from where did they come? The answer is that all these women formed a mosaic, a pattern quilt, a salad bowl, a melting pot of a few independent shtetls or villages or if not communities, than founding events only on the maternal side.

Does this mean that the Radhanites from Persia and Baghdad arrived in North Central or North East Europe and married local women? To find out, you'd have to test the mtDNA of Ashkenazi women and compare them to Jewish women in the Middle Eastern communities. Most of the women tested were not related to one another. Some were not related to sequences found today in the

Middle East and some were. So what you find is that mosaic, that salad bowl of isolated communities. With Europeans, if you want to see a bottleneck, you normally look at the extinctions that occurred during the Last Glacial Maximum ice age.

What many studies are finding about Jewish mtDNA is that 27% to 32% of Ashkenazim show mtDNA of K haplogroup, whereas only 2.6% of Ashkenazim have mtDNA U3. Compare that to 17% of non-Ashkenazi mtDNA in Iraqi Jews who have U3 mtDNA. What this shows is that the two Jewish groups have different founding mothers in those cases. Ashkenazi Jews also have mtDNA H and other types in different percentages. You can compare that with Russians who have mtDNA T haplogroup in 22% of the cases. One way to find out the origin of people today is to compare people's mtDNA for women and Y chromosomes for men. These are the ancestry markers, at least some of them.

You can also do racial percentages tests, but they won't reveal the origin as far as geography in ancient or medieval times. People move around and intermarry as well as create communities where they can remain isolated. For example, 51.4% of the Georgian Jewish women had a transition at mtDNA site 355, a particular mutation, when compared to the CRS, that standard Cambridge Reference Sequence against which mtDNA is compared to see mutations and differences from mtDNA that makes up about 46% of Europe and 6% of the Middle East. (I use Near East and Middle East interchangeably here). West Asia is not the Middle East. It's the Urals and N.W. Asia, including Turkey.

The Middle East is the Levant, Fertile Crescent, and Iran. North of the Middle East you have Western Asia, the steppes, the Urals. Beyond that is Central Asia. You'll find H mtDNA spread as a pan-European mtDNA haplogroup from Iceland to Bashkortostan. If you want to find the origins of Ashkenazi women, you can also go back further in time and read, "Tracing European Founder Lineages in the Near Eastern mtDNA Pool," Martin Richards, et al, *American Journal of Human Genetics*, 67: 1251-1276, 2000. If you're looking for the mother of all Europeans and Middle Eastern people, this article's for you.

What you'll find is that we ultimately all descend from the same single woman, and that lines are traced from the Near East to Europe and from Central Asia to Europe, but to get to Central Asia, you had to come either from the Middle East or back-migrate from Pakistan to Central Asia and across the steppes to Eastern Europe. Or you migrated from Southwestern Europe (Spain or S.W France) straight to North East Europe. The two peoples there met and intermarried, had children, and again, we come to the mosaic communities and melting pots.

Look further back at the genetic split occurring when people migrated from Africa to the Middle East and to Europe. The study found that Georgian Jewish women had very few founders. A late Paleolithic expansion took people from southwestern to northeastern Europe. See the article, "mtDNA Analysis Reveals a Major Late Paleolithic Population Expansion from Southwestern to Northeastern Europe," American Journal of Human Genetics 62:1137-1152, 1998.

You're trying to find out the history and geography of your Ashkenazi Jewish DNA, especially the mtDNA or matrilineal side—the mother's side. The mtDNA is how your ancestry is passed from one generation of daughters to the next. If you are male, you'll want to also look at the origins and migrations, the diversity in your Y chromosome. Women don't inherit a male Y chromosome, so for women, you'll be looking at your mtDNA, your matrilineal side and any other markers you are able to have tested. Perhaps you want to contemplate your most recent common ancestor or most ancient, long ago before nations had borders and places had religions with which we are familiar today.

What do the scientists tell us about the founders or origins of the Ashkenazi Jewish communities—especially about the sometimes nameless women in ancient or early medieval times and right up to the present? Research studies on the founding mothers of the Eastern European Jewish communities look at the mtDNA and report that the genetic patterns called the "modal haplotype" in Ashkenazi Jews has a frequency *similar to that of its host population* (9.0% vs.6.9%), according to page 1418 of the article, "Founding Mothers of Jewish Communities: Geographically Separated Jewish Groups Were Independently Founded by Very Few Female Ancestors," American Journal of Human Genetics. 70:1411-1420, 2002. The study was researched by Dr. Mark G. Thomas, Martin Richards, Michael E. Weale, Antonio Toroni, David B. Goldstein, et al—the rest of the team. Read the entire article for a scientific explanation.

For those with little science background, the point is for the non-Ashkenazi women the study recommended above found that there is a high frequency of particular mtDNA haplotypes in the Jewish populations, at least according to those people sampled. What that means is that a high number of Ashkenazi Jewish women have the same haplotype. They belong to the same mtDNA haplogroup in large numbers. In particular women of certain Jewish groups such as the Georgians, Moroccan Jews, and Bene Israel of India have high modal frequencies. This means that there were very few founding mothers.

When you compare male Y-chromosome and female mtDNA patterns, you see a striking contrast between the maternal and paternal genetic heritage of Jewish people. If you look at the Y chromosome, the study didn't find a consistent pattern of lower diversity in Jewish communities compared to non-Jewish host populations—the countries in which they lived for hundreds of years. This study contrasts with older studies showing many males having a Y chromosome relationship in 70 to 80 percent of the samples close to Lebanese and Syrian non-Jews and in other cases, a Y chromosome showing Eu 19, representative of 56 percent of Polish men and 60 percent of Hungarian men. What does this mean? Ten percent of Syrian men also have the Eu19 marker in their DNA.

With Jewish women, the pattern in the mtDNA is different than in the Jewish men. When you look at all Jewish communities, you see a significant lower mtDNA diversity than in any population with which the Jewish community is paired with other women's mtDNA markers. The finding shows that mistakes in associating any non Jewish population with any Jewish population would not influence the results of the latest study.

So what does this have to do with northern and Eastern European Jewish women, particularly those from northeastern Europe? When you look at female Israeli military recruits and see the Moroccan Jewish females with a cholesterol level of 128, and then compare them to the same age Eastern or Central European Jewish women with a cholesterol level of 162, eating similar foods, you wonder what is it in the mtDNA of the North African Jewish women that is keeping their cholesterol to such a low level and what health benefits will it bestow in later life? Were these women prepared for a tropical climate whereas the North East European Jewish women had bodies set up to withstand the famine and rigors of a cold Ice Age climate? Both live now in Israel, but their ancestors came from different areas. Are these women related genetically at some time in the past three thousand years or not?

Among the nine Jewish groups studied by Thomas's team, there are eight different mtDNA types that are modal with a very high frequency. What this means is that their mtDNA, their matrilineal ancestry by genetics was compared to the standard Cambridge Reference Sequence (CRS). A part of the DNA was looked at called the HVS-1 CRS, occurring which occurs at 46% in Europe and 6% in the Near East (Richards et al. 2000). What the study found was that all of the seven European and Near Eastern non-Jewish populations have the CRS as their modal haplotype. Moroccan Jews had 27% of their mtDNA following the CRS. Ashkenazim had 9% of their mtDNA following the CRS.

Only two of the nine Jewish populations studied have the CRS as their modal haplotype. What this means in non-science language is that of the nine groups of Jewish women, eight of those groups have different mtDNA types "that are modal" with a very high frequency. Another study showed a high level of K mtDNA in an Ashkenazim sample.

The modal haplotype is simply the most frequently occurring haplotype in a set of haplotypes. If you want to get mathematical, scientists often speak of the Cohen Modal Haplotype for Jews and most other Middle Eastern or Southern European peoples who carry the Cohen Modal Haplotype. Historians call the Cohen Modal Haplotype the "signature of the ancient Hebrew people". Historians call the Atlantic Modal Haplotype the signature of the Western European people.

Geneticists and mathematicians use the word 'modal' to describe the most often occurring value or set of values in a given distribution of values or set of values. It's a mathematical term from statistics. And genetics is the most statistical of the biosciences, especially the field of molecular genetics.

Compare the modal value to the median value which has half of the distribution above it and half below it. The average or mean value is the sum of the values divided by the number of values. The average can be seriously skewed by a few very large or very small values. So, to simplify, once again, the modal haplotype is the most frequently occurring haplotype in a set of haplotypes.

If you're now wondering what all this talk about Cohen Modal Haplotypes and Atlantic Modal Haplotypes in a man's Y chromosome (the chromosome that determines his gender) has to do with one's right to be a "real Jew" it's about as philosophical as asking who has the right to be a Christian, Moslem, Hindu or Buddhist. A "real Jew" is someone who takes Judaism seriously and practices it. The orthodox rabbis say a real Jew is a Jew who either has a mother who practices Judaism or a grandmother who practices Judaism or who converts to Judaism.

It doesn't matter whether your hair is blonde or black or whether you look like you just stepped out of Babylon wearing the mask of King Sargon, whether you have a honey complexion, melanin deposits, wavy dark hair, Mediterranean features, or whether you look like a freckled, silver-eyed, russet-tressed Sorb milk maid from medieval East Germany or Poland. Sorbs, the most Western of Slavs, have an ancient origin perhaps in West Asia or Europe. Slavs lived in Central Europe before the 6th century CE.

When they migrated south and east, there were Iranic-speaking people living there in ancient times. Don't let anyone call you intimidating names such as "a self-styled Jew." Some people like to label others with names to make them feel that they don't have a right to practice their religion or have a country of their own. If the tiny two percent of Y chromosome genes that make up your enriched tapestry of inheritance happen to be close to the Sorbs of Germany and Poland or anybody else, it shouldn't matter when you practice your religion or seek your core identity. These names get labeled on Ashkenazi Jews. Don't let anyone take away your core identity. No one in the world would call someone a self-styled Christian or any other religion.

How often do you find non-Jews calling the Ethiopian or Indian Jews self-styled? Not much. And these two groups have genes that differ from most Jewish groups outside their geographic areas. The fact is that the genes of most Jews cluster together when the Y chromosomes of the males are studied for ancestry, and most of these sequences also occur among non-Jewish males in the Middle East. However, the matrilineal lineages—the mtDNA of the Ashkenazi females often shows they are not related to one another and often not related to anyone in the Middle East.

Some of these women were descended from local women in communities founded independently who then merged together forming a mosaic. You can't say that they are all Slavs because they are not related to one another. If these women are not related to anyone in the Middle East, then to what geographic area are they related? It depends on their individual sequences. Until you look at the sequences of individuals, you won't know whether that person is related to sequences found in the Middle East.

The answer is a community that merged together from different origins has become a mosaic of diversity, and out of that diversity has come one people. Not as diverse as the United States, but nobody calls an American a self-styled American because of a diverse genetic ancestry. Still, looking at the women's lineages, you find a bottleneck in the past and today figures among Ashkenazi female lineages of 27 percent to 32 percent having K mtDNA. Georgian Jewish female lineages so very few female founders. The whole idea is to do detailed genetic analysis to find out how useful population-based gene mapping is in tracing the founding mothers of various Jewish communities.

The plethora of hate groups direct their diatribes to Jews who look the most like northern Europeans with labels like "Khazar warrior." Genetics so far has not compared the genes of "Khazar warriors" to the genes of Ashkenazim. What would they find? Two percent traces? Or would they more likely find a rich tapestry of diversity? Or certain groups within Ashkenazim such as Ashkenazi Levite males whose Y chromosomes are found in some cases to be close to modern Sorbs of East Germany and Western Poland?

Real Jews come in all colors and hair types. The fear factor is played upon when some group gets on the Internet and announces you don't have the right to be a Jew because you come from Northern Europe instead of Iraq, that your

European blood marks you as a self-styled Jew. Don't listen to that hogwash. There are no such things as self-styled Jews any more than there are self-styled Christians or any other religion. It's like saying you don't look Catholic because you aren't dressed like a Roman and speaking Latin so you must be part of the "barbaricum."

The argument comes from those who say that if you're Ashkenazi and have one drop of blood from a northern European, you have no right to call yourself a "real Jew" (and come to pray or live in Israel.) Your answer would be, show me that Palestinians or other Middle Eastern persons don't have one drop of Greek or Roman blood, as European as you can get, and therefore they become "self-styled Arabs" whose real language is Aramaic, reflecting the Greek, Roman, and Arab genes in their admixture. The point is we are all mixed mosaics of time, and the mixture often occurred long before there were any organized boundaries or religions. Back migration from Europe to the Middle East in prehistoric times is about 15% for all populations in that area. You are what you take seriously.

In the book, Genetic Diversity Among Jews by Batsheva Bonné-Temir and Avinoam Adam, Oxford University Press, 1992, chapter 4 is titled "Types of Mitochondrial DNA among Jews." Thirty-four different mtDNA types were found in their sample studied. The sample consisted of Ashkenazi Jews from Central and Eastern Europe, and Jews from Turkey, Iraq, Yemen, Habban, Morocco, and Ethiopia. The mtDNA was studied for heterogeneity, that is, genetic diversity. Researchers calculated the degree of heterogeneity.

Researchers look for patterns in several enzymes. Each pattern is given a different number. So a series of five numbers represents the gene fragment patterns that show up after the DNA sample is "digested" with the five enzymes, and this is what makes up one's mtDNA type. Then the researchers look at how the mtDNA types are distributed. Scientists then draw up tables to look at the degree of diversity each of the Jewish communities reveals. Researchers look at the number of people, the number of each person's mtDNA type. Then they look at how frequently each type appears. So the purpose is to study heterogeneity—that is diversity—and arrives at values of diversity for each Jewish community. The study of mtDNA is based on looking at the fragment patterns of the enzymes studied. In the chapter, five enzymes were studied based on their fragment patterns and presented in a table.

The researchers used five restriction enzymes, and 34 different mtDNA types were looked at from women from seven Jewish communities—Turkey, Ashkenazi (Central and East Europe), Iraq, Yemen, Habban, Morocco, and Ethiopia. Nine of the mtDNA types showed up in several of the Jewish communities—six in single communities and 19 were found in single individuals. The most common mtDNA type was found in 160 individuals out of 268 studied. The second most common type was found in four communities. These were 31 women from Ashkenazi, Iraqi, Yemenite, and Moroccan communities. The third cost common type was found only among the 12 Ethiopians studied.

Scientists looked for variability—diversity—heterogeneity. The overall variability of mtDNA types of these women was smaller among Jews than among the Caucasian population as a whole, Asians, Australians, or Africans. Variability is highest in Africans, lower in Jews. But some populations have mtDNA heterogeneity values even lower than Jews, such as Native Americans (Amerindians) or Finnish women.

The study found that the largest number of different mtDNA types (14 different types) were in Ashkenazi Jews. Ethiopian Jews were different from all the other Jewish groups. So there might have been an African contribution to the Ethiopian Jewish mtDNA pool. Ashkenazi mtDNA is close to Moroccan Jewish mtDNA. This was found in the study went another set of "six restriction enzymes was used." If you want to read further about the closeness of Moroccan Jewish mtDNA and Ashkenazi mtDNA, see the article "Mitochondrial DNA polymorphism in two communities of Jews," Tikochinski Y, Ritte U, Gross SR, Prager EM, Wilson, AC. (1991) American Journal of Human Genetics 48:129-136.

It's only by reading these articles in the various human genetics journals that you can follow the studies for a decade and see how new inroads are opening to finding out more about the founding mothers, the mtDNA of Ashkenazi women or other groups in which you are interested so you can perhaps get a handle on your own origins from any of these communities.

The question is how many mtDNA types in Ashkenazi or other groups of Jewish women are derived from or exclusive to one single Jewish community and how many are possibly derived from non-Jewish ancestries? What geneticists aren't able to estimate to the degree that most people want is the extent of admixture by looking at the markers. When the total human genome becomes affordable, individuals can see their own total genetic profiles and decide for themselves how rich their genetic heritage is from so many migrations, expansions and mixing or isolation.

Right now the extent of admixture can't be determined because scientists as well as science journalists like me cannot assume the lineage is or isn't the way it is due to extinctions of mtDNA in each community or due to admixture. All we know is that there are differences between communities. We don't know whether the differences are due to extinctions of lineages. So we can't estimate the extent of admixture...yet.

We do know that mtDNA in Ashkenazi women went through a bottleneck. We know that there is a striking high frequency of particular mtDNA haplotypes in Jewish populations, such as K mtDNA haplogroup showing up in 27%-32% of Ashkenazim. But when we look at K, it did come in Paleolithic times from the Middle East to Europe. mtDNA K is a clade of mtDNA U. That is, it is a branch off of U mtDNA and rather like its sister group.

Keep in mind that 17% of Iraqi Jewish mtDNA is U3. U mtDNA haplogroup has many branches such as U6 in northern Africa, U2 in Europe and in India, U7 in the Middle East, U5 in Europe, U4 in Europe and Armenia/Caucasus/Europe, and other branches off of U elsewhere from Iran (22% U mtDNA) and Central Asia to all over Western and Eastern Europe. H is the most frequent mtDNA haplogroup in Europe—about 46% and about 25% in the Middle East.

You have to look at the sequences and then at the tables to see where the sequences are present, more in Europe or more in the Middle East. The origin point is not where you have the highest number, but where the mtDNA sequences are the most diverse. Where you have the highest number tells you where most of the people migrated to and are living in today. Where the mtDNA sequences are the most diverse shows you where the mtDNA haplogroup origi-nated.

Haplogroup K mtDNA springs up 17,000 years ago in Northern Italy where Venice is located today. Then it migrates to the Alps. Jewish populations have high modal frequencies of mtDNA, particularly Moroccan Jews, the Bene Israel of India, and Georgian Jews. In "Founding Mothers of Jewish Communities: Geographically Separated Jewish Groups Were Independently Founded by Very Few Female Ancestors," Mark G. Thomas et al, *American Journal of Human Genetics*, 70:1411-1420, 2002, the findings were that most Jewish communities were founded by relatively few women.

So we know that the founding process was independent for different geo-graphic areas. We know that genetic input from the host country population was limited on the female side. You go back far enough and find out there was a major late Paleolithic population expansion from southwestern to northeastern Europe. Read the article, *"Genetic Affinities of Jewish Populations,"* Livshits G, Sokal RR, Kobyliansky E (1991) American Journal of Human Genetics, 49:131-146.

The mtDNA is inherited maternally and is gender-specific. Mothers pass it on to their daughters without much change for thousands of years. It's a way of looking back at ancestry, but it mutates over time, but slowly over thousands of years. Each mtDNA haplogroup originates from a single female founder who lived thousands of years ago, sometimes as long ago as 50,000 or 20,000 years.

When scientists look at the Y chromosome of males in Jewish communities, the Y chromosome shows diversity similar to that of neighboring populations, and shows no evidence of founder effects according to the article, "*Founding Mothers of Jewish Communities: Geographically Separated Jewish Groups Were Independently Founded by Very Few Female Ancestors*," Mark G. Thomas, et al, American Journal of Human Genetics 70:1411-1420, 2002.

The mtDNA is a female gender-specific X-chromosome microsatellite. The article also showed that the Georgian Jews have a female-specific founder that appears to have resulted in "elevated levels of linkage disequilibrium." What this all means is that when you compare Y chromosome and mtDNA diversity patterns you see a contrast between the male and female genetic heritage of the particular Jewish populations studied. Only with the Ashkenazi populations studied, the mtDNA diversity is more like the host country population. Also compare this with the studies of Y-chromosomes done by comparing different Jewish communities and comparing Jewish and non-Jewish Middle Eastern populations (Hammer et al 2000; Nebel et al 2000).

That means that males who have the Cohen Modal Haplotype will share a common ancestor with other males who have had ancestors in the Middle East in prehistoric times, possibly a common ancestor about 7,800 years ago. It also means the males descend from a common ancestor who lived about 3,000 years ago in the Middle East who could have been Aaron, the brother of Moses, and that the 3,000 year-old CMH existed in ancient Israel and is the signature of the ancient Hebrews. The CMH also is found in some Arabs. So the CMH had an ancestor still further back in time that both Jews, Arabs, and other Middle Eastern men shared 7,800 years ago. Genes have coalescence points, a time of origins that are thousands of years old.

What geneticists are not finding yet is an uncut umbilical cord going from ancient Israel to modern Jewish communities anywhere in the world. (de Lange 1984, p. 15; Encyclopedia Judaica 1972). The mobility of early Jewish communities is remarkable (Beinart 1992). You have conversions common in pre-Christian Rome and its empire. You have the Khazars and the Himyars adopting Judaism in ancient or medieval times. Jewishness is defined my maternal descent (Encyclopedia Judaica 1972.) So what happens when a Jewish man from the Middle East travels on trade business to Northern Europe and marries a local pagan woman? He converts her. By Jewish custom, when someone converts, the person is obliged to forget past the former religion, which usually was pagan in ancient and early medieval times.

From then on, the children are reared as Jews. Only one question: When a Jewish woman marries a Sorb man, if that's the case (as in Sorb Y chromosomes being close to some (not all) Ashkenazi male *Levite* Y chromosome genes), does he convert, and for emphasis, become not only a Jew, but a Levite, a servant, reader, or temple musician, of the Cohanim in the medieval synagogue? Historically, Levites were the tribe who lived in Egypt before the Exodus and who assisted the Cohens in the temple in ancient times. They came from the same family or tribe as the Cohens.

How all this relates to genomic nutrition is that your genes have a signature that express a need for certain types of food that are healthier for you than other types. The cultural and biological components of nutritional genomics also is expressed as part of the whole individual, culture, ethnicity, and above all, individividual genetic variability. You can't prescribe one type of food for one ethnic group, only for one individual. Some ethnic groups can tolerate some food better than other foods, but there is the exception with individuals as in some people not tolerating the lactase in milk and others tolerating it regardless of the ethnic group from which they originate. Genes are individual signatures within the cultural components of a society, and genes vary from person to person.

So, if the Khazars had their own Levites and Cohens, did the medieval Sorbs—at least the ones who may have converted? And did they merge later when the Jews were expelled from Germany into Poland in 1349? The main waves of migration into cities with large Jewish populations show that Jews migrated from Germany in 1012 CE to Poland near Bialystock and Grodno. By 1495 Lithuanian Jews also migrated to Bialystock Poland. But the founding families of Lithuania by tradition are said to have come from Babylonia in early medieval times.

You have the plague in 1347 followed by a 1348 migration of Jews gong from Eastern Germany to Lemberg and Temopol, and Hungarian Jews from 1349-1360 migrating to Temopol also. Earlier you have 1096-1192 migrations of Jews from North Eastern Germany going to Grodno, Poland, to Kalisz and Lodz in 1248, to central Poland and Krakow in 1159. Seems every time a plague broke out, the Jews were thrown out of some German town and went into Poland. You have Jews going from Austria to Temopol in 1421. Most of the movements are from Germany to lands East and North.

Then you have the Lithuanian migration of 1445 due Southeast to the Crimea. And you have a 1016 migration of Jews from the eastern cost of the

Black Sea near the Crimea in the Feodosiya area going to Kharkov. Interestingly both Byzantine Jews from this area as well as the Judaic Khazars who lived there might have gone to Khakov going from southeast to the north in Russia. At the same time the Jews of Kiev, a mixture of Byzantine, W. European, possibly Khazars and other Jewish communities living in Kiev moved into the Crimea in 1350. All this movement of communities focused around 1350, just after the plague receded.

You have to look not only at DNA but at rabbinic family histories, court Jews of medieval times and court Jews in the 16th century in the Germanic countries. In the Ashkenazi world, you have rabbinical dynasties as well as in the Sephardic or Mizrahi environment. Sometimes court Jews were members of rabbinical dynasties. According to an article that I highly recommend which you can find on the Web at: http://www.jewishgen.org/Rabbinic/journal/ashkenazic.htm titled, "Ashkenazi Rabbinic Families," by Dr. Neil Rosenstein, An Ashkenazi aristocratic dynasty held sway from medieval times to the end of the 16th century, especially in Austria and Germany. Court Jews married the children of other court Jews. According to Dr. Rosenstein's article, "Any Ashkenazi Jew living today has at least one in fifty chance of tracing back to a rabbinical dynastic lineage."

In the article's section on book printing, you realize the profound influence it had since the 15th century on scholarship in the Jewish communities. How would the cultural component of genes express themselves in the biological component of genes in any culture focusing on book printing for centuries? How would certain foods impact the genetic signatures of any ethnic group of people working in the field of book printing or copying scrolls for hundreds, even thousands of years—as as opposed to hunting, fishing, or farming outdoors? There would be less exposure to the agricultural soil, and TB in medieval times lurked in the soil.

How would the genes of farmers compare to the genes of hunters or urban dwellers in small, cramped ghetto quarters of Venice or Vienna? It's with the cultural and biological components of genes so well described by Dr. Scriver's book that today's researchers in nutritional genomics need to consider. What is there to learn about any particular ethnic group's cultural foods?

Which foods are intelligent for one person's genes and not for another person's genes in the same group? Food that is healthful for you may cause a metabolic problem for someone else. That person could be from your ethnic group or any other. Think of how a profession such as book printing might influence the way genes express themselves in relation to what kind of food is healthy or unhealthy for a person doing the same work day after day.

What if your genes were set up for hunting or farming and you sat all day in front of a computer and ate the same foods you would be eating if you walked with your sheep and goats six miles a day as the Masai in Africa have done with their cattle? How would your genes express themselves? Let's look at molecular genealogy.

In other resources on Jewish genealogy and history you see patterns linking book printing or scroll copying, family history, and genetics or DNA. That bridge is oral history, diaries, and family history journals.

When families talk about "the curse" from generation to generation, they may be referring to an inheritable disease or genetic event. For example, one Jewish family wrote of the "curse" on their family passed from father to son which turned out to be diabetes and glaucoma over generations. I'm using the example here if tracing Ashkenazi DNA, but you can trace your own ethnic group's ancestry not only to food habits or ancestry, but to how the food affected the person as occupations changed from agriculture to sitting all day at work. Did the food customs and type of foods remain the same?

How did the ethnic food habits influence the health of any ethnic group you look at or all ethnic groups? How is the individual within the ethnic group handling the particular food? Does the food contribute to health or cause problems? Did the food delay the onset of chronic disease or hasten it? Dentists might be interested in genetic profiles for gum disease or in studying the effects of polyphenols such as green tea extract or other antioxidants in prevention.

What foods are peculiar to your ethnic group and how have those foods influenced health of individuals rather than the group as a statistic? Let's look at tracing ethnic DNA. One example is the Ashkenazim. You can research your own group by tracing the DNA. Then apply what you discovered about DNA, ethnicity, and ancestry to nutritional genomics. What did your most recent common ancestor eat, and how did it influence his or her health? Let's look at the Ashkenazim as one example to follow through history. Did milk/lactase intolerance in a high percentage play a role? Or was a high dairy diet emphasized in spite of the symptoms that may have arose from lactase intolerance.

About thirty-two percent of North American Ashkenazim do tolerate milk well, and about 68 percent have some intolerance. How was milk treated to lessen the effects of lactase intolerance? What types of cheeses or yogurts were made from the milk? Greeks also have a similar level of lactase intolerance, about 66 percent. The many types of Greek cheeses and yogurt are well known.

In Southeast Asia, about 98 percent of the population cannot tolerate the lactase in milk. How have their foods evolved? To trace DNA for ancestry and then to put two and two together and apply it to another type of DNA testing for chronic disease risk and tailoring foods to one's genetic signature can be applied to any group.

Let's start with tracing the DNA of Ashkenazim for ancestry rather than for disease risk. It's the philosophy here than your own individual genes will show you what risks you have when you look at the results of your "nutritional genomics" genetic profile, a different kind of DNA test from the test for ancient ancestry. You have two kinds of DNA testing—one for chronic disease risk of SNPs so you can find out what foods and supplements will lessen the chance of getting any chronic problems your genes put you at risk for, and another DNA test for ancestry that does not test any areas of your genome revealing disease risks.

The DNA test for ancestry tests only your maternal line, your mtDNA or Y-chromosome, paternal line (in males). All it shows is where your deep, ancient maternal or paternal original ancestor might have come from before migrating elsewhere. You'll get your markers for ancestry. You can use that information to look for DNA matches of mtDNA or Y-chromosome, but the mtDNA or Y-chromosome represents only two to three percent of your entire genome.

You can also take a racial percentages test. Again, these types of DNA tests do not test for any diseases. You'd need to take another type of DNA test of SNPs to see what chronic disease risks you might have for the purpose of tailoring foods to prevent or delay those chronic illnesses. Again, it shows the risk, not that you have the disease. Use that kind of test to tell your heirs that you carry those genes and could pass on the same risks. However, your heirs would have to have their own tests to see whether or not they have inherited the same risks you have.

Everyone's genes are different, but at the same times, inheritable diseases do sometimes get inherited. Only testing will let you know your risks. Testing for DNA ancestry is fun, because you don't have to think about disease, only deep ancestry. You'll get a printout of where people with your DNA live today with some DNA companies or you can search online databases for matches. This does not mean they are related to you now, but that you may have shared a common single ancestor in the last ten or twenty thousand years, and sometimes a lot closer to the present.

Tracing Jewish or any other ethnic group's DNA is like investigating a mystery with many clues that may be found in records of journals and books from the time book printing began to influence the world of Jewish scholarship, writing, and study. The rabbinic world was where writing was taking place in the past centuries. Families tended to form particular cliques or groups within some Jewish communities to ensure certain positions for their heirs.

So tracing the DNA of your most recent common ancestor 250 years ago could point to something fascinating in your background possibly linked to involvement of Jews with book printing in Europe or elsewhere. Before book printing there was scroll copying with ink and quiver. You have to go back further to look at DNA and Jewish migrations, back to the 14th century.

The years 1347-1350 meant huge migrations from West to Eastern lands for Jews. It also coincided with the bubonic plague. Jews were expelled because the people didn't know the plague came from fleas piggybacking on rats. Many blamed Jews for poisoning the wells. So large migrations or expansions of Jews coming out of Germany, Hungary, Lithuania, and Central Europe moved east into Poland and parts of Russia.

About one third of the Jewish population was said to have succumbed to the bubonic plague. There were also massacres, problems with the crusaders, and most communities were on the move. Jews from the West went east and those from the east went northeast. Nobody was going West from Russia or Poland back into Germany between the 11th century and the 15th, at least by the looks of the arrows on the migration map of Jewish waves of migration, and these were only the main waves of migration to cities with large Jewish communities.

So we know that the mtDNA of Ashkenazi women is more like the host country population as far as diversity. However, we can't assume it's because of admixture because the bottleneck in the past that occurred from possibly extinctions could have created the situation of a lesser number of mtDNA types in the founding mothers of the Ashkenazi European Jewish communities. What we can assume right now until other evidence comes forth is that the communities were made up of a mosaic of different communities of Jews from different geographic areas that eventually merged somewhat.

We can't assume that just because Sephardic and Ashkenazi Jews spoke different languages and had different prayer rituals that they didn't marry. They did. Sephardic Jews settled in Germany, Holland, and other European countries. Some married only with other Sephardim and a few married with Ashkenazim.

After the end of the Spanish Inquisition, many Sephardic Jews migrated to Livorno, Italy, and from there, sailed to Aleppo, Syria where two groups emerged, the Signoreem from Spain and Italy and the Mustarabim, living in Syria for thousands of years. Ashkenazim moved to Turkey, Bulgaria, Serbia, Romania as well as Polish Jews moved in the 18th century to Romania where both Ashkenazim and Sephardim lived in the same country. The extent to which they communicated with one another varies from country to country.

Even in medieval times, Sephardic travelers from Spain visited Russia, and the Balkans were under the Ottoman Turks for hundreds of years, then under the Russians. Jews moved around. Radhanites peddled their wares in the Middle East and in Poland in medieval times.

You have groups such as the Bene Israel of India whose mtDNA is similar to Hindu people of India. Yet the men they married could have come from the persecutions of Epiphanus in 175 BCE. The Bene Israel have a local tradition that claims they are descendants of refugees from Antoichus Epiphanus. Over two thousand years, they might have taken local wives from that area of India where they live.

You'd have to test the Y chromosomes to see whether the males are linked to ancient Israelis or at least link to other Jewish communities. The Beta Israel (Ethiopian Jews) show mtDNA related to other non-Jewish Ethiopians. Again, you'd have to test their husbands or fathers to see how they relate in the scheme of genes. The point is anyone can become Jewish at any point in time. Jews come in all colors and mtDNA haplogroups or Y-chromosomes. What you're looking for is heterogeneity when you look at Jewish genes. You're looking for diversity, variability in the various communities. Ethiopian Jewish tradition says they are descended from Jewish noblemen who came with Menelik on his journey from Israel/Judea in the early years of the first millennium BCE.

So far few if any have compared ancient Israelite bones to any particular group of Jews living today, but Iraqi Jews are said to be closest to ancient Israelites in both mtDNA and Y chromosomes. There are Jews in Persia who are a branch of Iraqi Jews. Then you have Persian Jews who joined the community of Bukharan Jews in the third to seventh century of the Common Era. Jews of Central Asia, Persian Jews, and Iraqi Jews all have slightly different mtDNA sequences. Georgian Jews have mtDNA types that show they may have come from very few local women. For example, a Jewish community thrived in Tbilisi in the 4th century CE.

What's missing from articles in English in various genetic journals are studies that compare the mtDNA or Y chromosomes in the bones or teeth from Khazari burial sites in Russia or Daghestan with Ashkenazi Jews, with modern non-Jews today in various communities nearby what was once Khazaria, and with Jews from non-Ashkenazi communities. Nutritional genomics and nutritional anthropology or archaeogenetics can trace foods that a population ate by looking at the lipids and a variety of foods left on the pottery.

Since so much literature abounds trying to related millions of Polish and Russian Jews to descendants of Khazar warriors, a Turkic tribe from Central Asia whose homeland actually was in what today is Daghestan—and also those lands where the Volga meets the Caspian. Cossacks presently occupy the lands that were in medieval times inhabited by Khazars (Khazari, Kosarin).

Operas have been written about Khazar princes in love with Russian maidens. There is a Viking-like aura about the Khazar warrior whose nobles, perhaps about 4000 convert to Judaism in early medieval times. The Khazar Empire ruled Slavic peoples who may or may have not converted to Judaism. Today, the language of Chuvashia is most closely related to the medieval Khazari language. Chuvashia is a country of mainly Christian peoples (Eastern Orthodox) called the Chuvash, a Turkic-derived Christian population.

Since Ashkenazim speak Yiddish, looking to the roots of Yiddish we find Sorbian (not Serbian) also known as Lusatian and called Wendish by the Germans (their word for Slavs) at the root of 13th century-derived Yiddish. The Sorbs lived in Prussia and in the areas in Eastern Germany near Poland and in Western Poland. In Yiddish, German words quickly replace the Slavic words spoken by Sorbs.

They are the most Western and most little known Slavic peoples living in a country called Lusatia. According to the Columbia Electronic Encyclopedia, at http://www.factmonster.com/ce6/world/A0830646.html Sandy and forested or hilly and agricultural Lusatia is located in a region of East Germany and Southwest Poland. Jews lived in Germany and Poland, but did they live among the Sorbs?

The question for geneticists and historians is whether the Ashkenazim lived in medieval times in small communities in the vicinity of Lusatia, land of the Western Slavic Sorbs? The area in question extends north from the Lusatian Mts., at the Czech border, and west from the Oder River. Did Jews live in the sandy northern or hilly, agricultural southern Lusatia?

The particular area in question is where the Lusatian Neisse separates E Germany and SW Poland. The Sorbs were farmers and foresters, mostly raising livestock. Did the Jews come there to work the lignite mines, textile mills, and glass-making factories in medieval times? The towns in question would be Bautzen, Cottbus, Görlitz, Zagan, and Zittau—the primary Sorb towns. The Lusatians are descended from the Slavic Wends. Wendish is still spoken in the Spree Forest. Traditional Wendish traditional dress and customs are preserved. The word Wendish is what the Germans called the medieval Sorbs.

The Ashkenazi communities left Southern Italy in the 4th century and by the eighth century migrated to France and the Rhineland. Did they take Sorb/Wend brides? Did the Sorb men convert to Judaism? Was it the Sorb men, not the Khazars who mixed more into the Ashkenazi population? Or is all this legend?

By the 10th century, German and some Polish Jewish communities were becoming known for their rabbinical scholarship. We know Jewish communities existed in the Rhineland and France by names such as Speyer/Shapiro, Luria for the Loire valley, and other Ashkenazic rabbinic families in the areas. When the Jews may have first met the Sorbs, the Sorbs were not yet Christianized by the Germans. Did the Jews taken pagan brides and convert them? Or did the pagan men convert to Judaism before the Germans colonized the areas of these Western Slavs and converted the remaining pagan inhabitants to Christianity?

The Sorb/Lusatian/Wend region was colonized by the Germans beginning in the 10th century. German rules constituted the area into the margraviates of Upper and Lower Lusatia. Both margraviates changed hands frequently among Saxony, Bohemia, and Brandenburg. In 1346 several towns of the region formed the Lusatian League and preserved considerable independence. By 1349, Jews migrated in large numbers eastward to found the shtetls and villages of their communities in Poland.

Back in Lusatia, Under the Treaty of Prague (1635) all of Lusatia passed to Saxony. The Congress of Vienna awarded (1815) Lower Lusatia and a large part of Upper Lusatia to Prussia. After World War II the Lusatian Wends (or Sorbs, as they are also called) sought unsuccessfully to obtain national recognition. Since Sorbs spoke a language somehow related to Yiddish, are the two peoples linked genetically? See the article, *"The Origins Of Ashkenazic Levites: Many Ashkenazic Levites Probably Have A Paternal Descent From East Europeans Or West Asians"* Bradman, N1, Rosengarten, D and Skorecki, K21, The Centre for Genetic Anthropology, Departments of Biology and Anthropology, University College London, London, UK. And 2 Bruce Rappaport Faculty of Medicine and Research Institute, Technion, Haifa 31096, Israel and Rambam Medical Center, Haifa 31096, Israel. (See the Bradman Index at http://dna6. com/abstracts/bradman.htm which is online as part of an index to a list of abstracts and posters at http://dna6.com/abstracts/index.htm also see at: http:// dna6.com/abstracts/filon.htm the article, *"An Unexpectedly High Frequency Of (-Thalassemia Carriers In Ashkenazi Jews,"* Filon, D, Jackson, N, Oron-Karni, V, Oppenheim, A and Rund, D. The Hebrew University and Hadassah Hospital, Jerusalem, Israel.

In the article, frequencies of high resolution Y chromosome haplotypes characterised by 11 biallelic polymorphisms and six microsatellites were analyzed within the context of their genealogical relationships to investigate the paternal origins of the Ashkenazi Levites. When comparing Sephardic and Ashkenazi Levite datasets, the results of the study suggests that, "unlike the Jewish priesthood (Cohanim), there is no evidence for a shared patrilineal origin for this caste across Jewish communities."

So, are Ashkenazi Levites from Eastern Europe really sons of Sorbs? "In the research, Levite haplotype distributions were compared with distributions in Israelite Jews and candidate source populations (north Germans and two groups of Slavonic language speakers). The Ashkenazic Levites were most similar to the Sorbians, the most westerly Slavonic speaking group," the article reports.

Linguists studying the grammar of Sorbian and of Yiddish, the language of the Ashkenazim propose that on grammatical grounds, for Sorbian to have contributed to Yiddish—the language of Ashkenazic Jews. According to the article, "Comparisons of the Ashkenazic Levite dataset with the other groups studied suggest that Y chromosome haplotypes, present at high frequency in Ashkenazic Levites, are most likely to have an east European or west Asian origin and not to have originated in the Middle East."

So just when we think, well, the Levites of Eastern Europe have Y chromosomes that show male (paternal) ancestry down through the millenniums, coming from the Slavic Sorbs of East Germany and Western Poland, when did this occur? How far back in early medieval times did they convert? Was it before the Germans came to the Sorb lands to convert the pagan Slavs to Christianity? Then where geographically did the Jewish mothers originate and in what century?

In non-Ashkenazic communities, there were few female founders of Jewish communities. We know there was a bottleneck in mtDNA in the Ashkenazic communities. Yet, there is somewhat more diversity in the Ashkenazic communities in mtDNA than in the non-Ashkenazic. Only not too much diversity with 27% to 32% of Ashkenazic mtDNA being K mtDNA haplogroup and 9.0% being H haplogroup and following the CRS.

According to the Jewish Virtual Library at: http://www.us-israel.org/jsource/Judaism/Ashkenazim.html, in 1182, "Jews were expelled from France. Ashkenazi Jews continued to build communities in Germany until they faced riots and massacres in the 1200s and 1300s. Some Jews moved to Sephardi Spain while others set up Ashkenazi communities in Poland." The center of Ashkenazi Jewry shifted to Poland, Lithuania, Bohemia and Moravia in the beginning.

The center of Ashkenazi Jewry shifted to Poland, Lithuania, Bohemia and Moravia in the beginning of the 16th century. Jews were for the first time concentrated in Eastern Europe instead of Western Europe." So if your mtDNA has a center in Northern France or other areas of Western Europe, but your relatively recent ancestors came from Eastern Europe, perhaps the matrilineal line came into the Ashkenazi community when Jews lived in France in the 12th century or earlier. In the Jewish Virtual Library, it notes that, "In the 1500-1600s, Polish Jewry grew to be the largest Jewish community in the diaspora." Yet in medieval times, "Centers of rabbinic scholarship appeared in the tenth century in Mainz and Worms in the Rhineland and in Troyes and Sens in France."

As a genealogist using DNA to find matches, think about the possibility that if you're geographic center is in northeastern France, perhaps this is when you're direct matrilineal line came into the picture. For Y-chromosomes, it would signal your paternal line in men. Or perhaps it didn't. Can you really know for sure? Technology in the future will be more precise sequence DNA than it is currently. Could the fewer mtDNA haplogroups be due to extinctions? For non-Ashkenazic communities, there were at least eight founders. How many female founders were there in Ashkenazic communities? These are questions only research will answer as more unpublished information and research becomes available.

According to another article on the high number of thassalemia (a type of severe anemia) carriers in Ashkenazic Jews, *"An Unexpectedly High Frequency Of (-Thalassemia Carriers In Ashkenazi Jews,"* Ashkenazi Jews resided for many centuries in a region in which Falciparum malaria was not prevalent. You'd think Eastern European Jews wouldn't have a significant frequency of (-thalassemia. Yet they do. So do Sephardic and Middle Eastern Jews. And also some non-Jewish Poles—so far 52 cases reported in Christian Poles.

So did the Ashkenazi Jews inherit their mutation for getting this type of anemia from their Mediterranean ancestors in ancient times, or from medieval and modern Christian Poles who are popping up with it, or at least reporting it? Or did they get it from intermarriage with traveling Sephardic or Middle Eastern Jews, such as the Iraqi and Persian Radhanites of early medieval times who came to trade what they peddled in Poland? According to the article, "Surprisingly, a genetic study of thalassemia on a referral population of various Israeli ethnic groups suggested a high frequency of (-thalassemia in Ashkenazi Jews."

According to the article, "Two nearly identical (-globin genes, (2 and (1 are present at the tip of chromosome 16 within a duplicated region. This arrangement leads to unequal crossing over and recombination between the homologous chromosomes. The most common product is a single (-globin gene deletion of 3.7kb, -(3.7, resulting in mild microcytic anemia. This allele was found in 87% of the (-thalassemia genes of Ashkenazi Jews."

Many genetic research studies find that they negate a hypothesis rather than prove it. The article reads, "Out of 151 samples only one was found to carry the (anti3.7 allele, negating the hypothesis; b. founder effect and genetic drift; c. selective advantage against P. vivax malaria; d. non-malarial selection to a linked factor. As there are a number of developmentally important genes linked to the (-globin cluster, the last explanation is an attractive possibility."

Basically, geneticists look for attractive possibilities when they study genes. The scientific method usually starts with a hypothesis, and the research tries to find out whether the hypothesis is negated or whether an explanation is an attractive possibility. Can genes prove anything? Only that there are results in print in the scientific journals waiting until the next researcher brings in new information. What can we know about genes?

If you read the book, **Genetic Diversity Among Jews,** you'll find chapters on the major demographic trends of world Jewry, on types of mtDNA among Jews, but the book has a subtitle: "Diseases and Markers at the DNA Level." So I highly recommend this book on any books that come after it on the subject of discussing the genetic details of inheritable diseases such as the chapter, "Tay-Sachs Disease Mutations Among Moroccan Jews." It's not just an Ashkenazi mutation. Different mutations for Tay-Sachs show up in Moroccan Jews and still other mutations in the French Canadians. For example, more than one mutation underlies TSD in Moroccan Jews.

You have a population of approximately 1.5 million Ashkenazim compared to about 500,000 Moroccan Jews living in Israel. Intermarriage between Ashkenazim and Moroccan Jews is common in Israel. The studies on the book were done on people living in Israel. In Ashkenazim three mutations account for nearly al the genes that cause TSD. There's also a milder variant that strikes adults, a type of gagliosidosis. Moroccan Jews have another TSD mutation. The Moroccan Jews came from a variety of cities—Casablanca, Marrakech, Rabat, and Res. Geneticists would have to trace their prior origins back and to relate their mutation with the geography and history of the people in their ancestral towns. It's difficult to do because the carriers come from large cities and populations that have intermarried for centuries.

So there are many other chapters such as "Genetic Complexities of Inflammataory Bowel Disease and Its Distribution among the Jewish People." Each of the chapters is by different authors. So the book is a wealth of information about not only the history and geography of Jewish genes, but mostly about the distribution of disease among Jewish populations from all over the world, but with the populations of this mosaic now living in Israel and studied there.

You find conclusions such as the less common Tay Sachs mutation could not have been kept in the Ashkenazi population by solely drift. So you may have to infer some advantage for carriers of the Tay-Sachs mutation. What was this advantage? The field of molecular genetics is studying this.

If you consider a career in this area, you would be thinking about admixture rates, population size, and a variety of selective mechanisms. The gene you pinpoint, the heterozygote advantage is part of ongoing science in looking at the whole genome. So as geneticists study these details, genealogists look at oral history and genealogy in a new way that combines DNA testing with looking at how diseases or mutations for disease are inherited. On the other hand, DNA testing can help you match your DNA to your most recent common ancestor.

If you have your DNA tested and find a match, perhaps this person shared a common ancestor with you, either male or female about 250 years ago, more or less. That's why databases used in genealogy and for DNA analysis find a common meeting point between those trained in science and those trained in family history.

So what is the conclusion here? We are a mosaic of different communities that have eventually merged. Let's take a look at the Middle Eastern Jews. Since Iraqi Jews are supposed to be the closest you can get genetically to the ancient Israelites, they have been living in Iraq since the destruction of the first temple in 586 BCE. Since late antiquity, you have Jews living in Morocco.

Not all of them came there from Spain in the 15th century. Jews from Yemen dwelled in the ancient Jewish kingdom of Himyar which is in Yemen. There were several Jewish kingdoms in ancient and medieval times—countries ruled by a Jewish king or queen. Examples include Himyar in Yemen, a Jewish Queen in northern Iraq, and Khazaria, where the Volga meets the Caspian. There were Jewish kingdoms in the Arabian Peninsula.

 See the Jewish Kingdoms of Arabia (from 4th to 7th centuries) Web site at http://www.eretzyisroel.org/~jkatz/arabia.html. The second paragraph in the article "The Jewish Kingdoms of Arabia 390-626 CE" at the Web site reads: "According to Moslem tradition, conversion to Judaism started under Abu Karib Asad (ruled 390-420), who became a Jew himself and propagated his new faith among his subjects. Arabic sources expressly state that Judaism became widely spread among Bedouin tribes of Southern Arabia and that Jewish converts also found with the Hamdan, a North Yemenite tribe. This time, many of the upper strata of society embraced the Jewish faith. The position of Judaism in Yemen reached its zenith under Dhu NuwAs."

Interestingly, the Habban (city in Yemen) Jewish community seems to have come from one single maternal origin. That means a single female actually appears in the study to be the founder of the Habban Jewish community. You

can view photos of the town of Habban in Yemen at the Web site at http://
www.alovelyworld.com/webyemen/htmgb/habban.htm. When the Jews of Habban left Yemen, they carried with them their special skills in working silver which
had made the town famous for its silver crafts. Most of the Habbanite Jews settled in Israel.

Referring back to the book by Batsheva Bonné-Temir, et al. 1992 book,
Genetic Diversity Among Jews, chapter four titled, "*Types of Mitochondrial
DNA Among Jews,*" what the mtDNA study found focused on diversity. While
there was no significant difference between the *heterogeneity*—the variability—in
the Moroccan Jewish community and those from the Central and East European
(Ashkenazi), Iraqi, and Yemenite Jewish communities, (that is the diversity pat-
terns were similar), there were significant differences between the Ashkenazim
and the Iraqi Jews. And the differences between the Ashkenazim and the
Yemenite Jews were statistically significant also. Read the chapter on "Nonisulin
Dependent (Type II) Diabetes Mellitus in Jews."

I've read in popular magazines that in medieval times in Central and North
Eastern Europe, Jews got diabetes and mental problems, and Christians got
tuberculosis. You didn't find too many Jews with TB. Was it the custom of washing the hands three times before eating? Tuberculosis bacteria was often found in
the soil in medieval times, and not too many Jews were farmers. Most were ban-
ished to small apartments in Jewish ghettos where they did close work such as
copying books, minting coins, money changing, or sewing and diamond cutting.

Think about type II diabetes—possibly caused by lack of exercise and eating
too many carbohydrates? Or mental illness—caused by oppression and living in
tiny ghetto apartments instead of on spacious farms? When the plague broke out
in 1347, Jews in the Rhineland were blamed and had to leave Germany. By 1349
they were streaming into Poland to join the small earlier communities founded
by Jewish Rhadanite merchants from Persia and Byzantine Jews who came from
the East at a time when Aramaic was still spoken by Jews in Baghdad.

This was at a time when the Yiddish language seemed to have developed out
of Sorbian, but the Sorbians had no written language, and the Jews had Hebrew.
Did the Sorbs, fleeing Germanization join up with the Jews? Only the genes will
tell. Now you have DNA tests of Y chromosomes on Ashkenazic Levite males
showing some of them have Y-chromosomes close to what the modern Sorbs
have. What does this all mean? A richer tapestry, a mosaic quilt of communities
merging?

How do the Ashkenazim relate to the Iraqi Jewish community? Prior studies
related Ashkenazic males to Iraqi Jewish males. Looking at the mtDNA of the

women of Ashkenazic and Iraqi Jewish communities, you find a lot of U3 mtDNA, about 17 percent in the Iraqi Jewish mtDNA—the matrilineal lines. And you find about 27 percent to 32 percent of K mtDNA in the Ashkenazi Jewish female—matrilineal lineages.

In the study, "Types of Mitochondrial DNA Among Jews," Additionally, there were significant differences between the Iraqi Jewish community and the Yemenite Jewish community. You need to keep in mind you're looking at a sample that tested 75 Ashkenazim, 8 Turkish Jews, 26 Iraqi Jews, 65 Yemenite Jews, 17 Habbanite Jews, 22 Moroccan Jews, and 55 Ethiopian Jews. That's 268 people in the sample. The mtDNA is what was studied for diversity—heterogeneity—variability in that small sample. So you have to research how significant a small sample of 268 is before you reach a conclusion that involves a whole population.

Interestingly, past studies of Y chromosomes found only in men, showed much more similarity between the various Jewish groups. Look at the article "High-Resolution Y Chromosome Haplotypes of Israeli and Palestianian Arabs Reveal Geographic Substructure and Substantial Overlap with Haplotypes of Jews," by Almut Nebel, Ariella Oppenheim, Mark G. Thomas, et al, appearing in the publication *Human Genetics* (2000) 107:630-641. Y-chromosome variation in the Israeli and Palestinian Arabs was compared to the same in Ashkenazi and Sephardic Jews. Then all of these Y chromosomes were compared to the Y-chromosomes of North Welsh males.

The findings: The Y chromosome distribution in the Arabs and Jews was similar—but not identical. Then the study went further to look at the haplotype level that is determined by certain markers. Haplotypes of Y chromosomes are determined by what scientists call "binary" and "microsatellite" markers. These genetic markers showed a pattern in finer detail. So when the scientists looked deeper to reveal more detailed markers, what they found was that there is a common pool for a huge portion of the Jewish and Arab Y chromosomes found only in males. This common pool suggests a recent common ancestry. By recent we are talking about several thousand years.

What the researchers saw was that the two "modal haplotypes" in the Israeli and Palestinian Arabs were closely related to the most frequent haplotype of Jews which is called the Cohen Modal Haplotype or CMH. The only problem is that not all Jews have the Cohen Modal Haplotype. And you don't have to be named Cohen to have one. Neither do you have to be a Cohen, a direct descendant paternally of the ancient Hebrew priests of the temple to have the CMH genetic marker. Yet the CMH is said to be the genetic marker of the ancient Hebrew people as it originated around 3,000 years ago at a time when Moses's brother, Aaron become the first Cohen or high priest of Israel. The marker has a common ancestor in the Middle East that relates Jews to Arabs and other middle eastern peoples dating back to a paternal ancestor or group of ancestors who lived around 7,800 years ago.

In the article, however, the study found that the Israeli and Palestinian Arab clade that includes the two Arab modal haplotypes that makes up 32 percent of Arab chromosomes isn't found much among Jews. In fact it's found at a low frequency among Jews. So there is divergence or admixture from other communities in which the Jews lived, but how much is the question? Nevertheless, other studies show that Arabs and Jews share about 18 percent of there gene markers but not too much of the distinctive Arab clade shows up in Jews.

On the other hand, the Cohen Model Haplotype and another modal haplotype in the Israeli and Palestinian Arabs were closely related to the Cohen Modal Haplotype in the Jews. Will scientists be able to go further and show that a sizable number of Palestianian Arabs once were Jews in ancient times? We already know some Jews once were Middle Easterners before Judaism existed. Again, we see how closely people were related in ancient times and how diversity began to spread, even slightly, when people migrated to other communities.

In the article, it reads, "Based on two Y chromosome RFLP markers it was suggested that Ashkenazi and Sephardic Jews are more closely related to Arabs from Lebanon than to Czechosovakinas (Santachiara-Benerecette et al. 1993.)" Also, the article reveals that "a recent survey of 18 binary Y-specific polymorphisms showed that Y chromosome haplotypes of Middle Eastern non-Jewish populations are almost indistinguishable from those of Jews (Hammer et al. 2000)."

The study also found a significant difference between Sephardic Jews and Arabs due to a higher frequency of haplogroup 2 in the Y chromosome of Sephardic Jews. Yet there was no significant differences found between Arabs and Ashkenazi Jews nor between the communities of Sephardic and Ashkenazic Jews. Yet Arabs, Ashkenazi Jews, and Sephardic Jews all were distinguished from the Welsh samples.

The major portion of Y chromosomes of Arabs and Jews belonged to haplogroup 1. Most of the Welsh Y chromosomes belonged to haplogroup 2. Yet the frequency of haplogroup 2 chromosomes in Arabs was lower than that in Sephardic Jews and slightly, but not significantly lower than that in Ashkenazim. However, Ashkenazim were closer to the Welsh than were Ashkenazim to Arabs. Sephardim were closer to Arabs than to Welsh.

So what does this say for Ashkenazi men? Does it say they have northern European admixture or genetic drift? You can't make this assumption because two Ashkenazi individuals carried the most common Welsh haplotype, and that finding can't be based on assumptions because if two men carry the most common Welsh haplotype, it's also the most common haplotype for most other European countries such as England and Norway or Frysland, all in northern Europe.

In the study three Arab individuals from a variety of areas carried haplotype 27 which is the Cohen Modal Haplotype, the signature of the ancient Hebrew people and also found in some Arabs. None of the Welsh in the sample carried the CMH. Neither of two modal haplotypes found in Arabs showed up in the Jewish sample.

When the Romans conquered Palestine and built their villas and cities there, you get gene inflow from the Romans and some Greeks. Interestingly, in ancient times numbers of Palestinians and Jews became Christian by the fifth century of the Common Era. (Bachi 1974). Then in the first millennium of the Common Era, immigration of various Arab tribes into Palestine increased with a huge wave coming in during the seventh century CE. Some of the Christianized Jews in Palestine along with the Christian Arabs converted to Islam at that time. (Saban 1971; McGraw Donner 1981).

Then you get more people moving into Palestine in later times such as Turks and Crusaders. You get recent migrations and gene flow from Europe. So Palestinian Arabs vary genetically as well as the Jews. What Palestinians share with Jews are linguistic backgrounds as Hebrew and Arabic share common origins, and geographic origins at their roots.

Researchers are studying their genetic relationships. Okay, there are genetic affinities between Arabs and Jews. Look at articles studying mtDNA (Bonné-Temir et al 1986; Ritte et al. 1993) Bishara et al 1997. There are affinities and there are differences. The point is Ashkenazi and Sephardic Jews are more closely related to Arabs from Lebanon than to Czechs and Y chromosome hapltoytpes of Middle Eastern non-Jews populations are almost indistinguishable from those of Jews (Hammer et al. 2000). But what about the founding mothers of the Jewish communities? What about Ashkenazi women? Are they Middle Eastern? Slavic? Germanic? Or? The answer is they form a mosaic of different communities formed from a bottleneck in ancient and medieval times.

What is known is that numerous people converted to Judaism at various times in ancient and medieval periods and in different countries varying from Southern Italy and Greece to Northern, Central and Eastern Europe. On the other hand, some people converted from Judaism to Christianity in ancient and medieval times to escape oppression and now exist in the Europe population with those same ancestral Y chromosomes or mtDNA.

Now the question remains, did more people convert to Judaism or convert to other religions in historic times? Whatever the answer is, it will show up when you compare gene markers of Jews and non-Jews in any European population or in the Middle East, considering back-migration from Europe to the Middle East taking place from prehistoric times to the present. During and right after the end of the Ice Age people in several waves back-migrated to the Middle East from Europe, which shows about 15 percent of the Middle Eastern peoples consist of individuals who migrated there from Paleolithic Europe.

Mutations are any heritable change in a DNA sequence. Polymorphisms are differences in DNA sequences among individuals that may underlie differences in ancestry or differences in health. Genetic variations occurring in more than one percent of a population would be considered useful polymorphisms for genetic linkage analysis. To study mutations or genetic variations in DNA sequences, scientists put your DNA through a polymerase chain reaction (PCR) test. It's a method for amplifying a DNA base sequence. The researchers usually use a heat-stable polymerase and two 20-base primers, one complementary to the (+) strand at one end of the sequence to be amplified and one complementary to the (-) strand at the opposite end.

So you are copying DNA. You're synthesizing DNA strands. These newly synthesized DNA strands will serve as additional templates for the same primer sequences. You go through successive rounds of primer annealing and do things like strand elongation and dissociation. What's produced are rapid and highly specific amplification of your desired sequence of DNA. Then you can use your PCR that is your polymerase chain reaction test to detect the existence of the defined sequence in your DNA sample. Okay, now you understand what happens when DNA is tested to find certain sequences.

Polymerase DNA or RNA also refers to an enzyme that catalyzes or brings together the synthesis of nucleic acids on pre-existing nucleic acid templates. You assemble DNA from deoxyribonnucleotides and you assemble RNA from ribonucleotides. So that's how it all comes together to see the human genome or part of it.

Numerous Jewish men, including Ashkenazim are genetically indistinguishable from Arab Palestinians, and some share the Cohen Modal Haplotype (CMH). This is a subset of an insular group such as Jews, Finns, Sardinians, and Basques. The DNA tests shows up the genetic pedigrees going back around 3,000 years to the Exodus from Egypt. Europeans in general are mixed and intermarried for thousands of years so that's it much more difficult to tell a Swede from a Russian or any one country from another in Europe. On the other hand, a study of the DNA of the Ashkenazi Male Levites revealed that some of then had Y chromosome DNA markers showing descent at least paternally, from East Europeans or West Asians.

So what can geneticists say about Levite Ashkenazi males? Are they the only Jewish males in Eastern Europe to possibly be descended from fathers who were Sorbs? What about non-Levite males? We know a percentage of the Cohens really do inherit the Cohen Modal Haplotype (CMH) that links them to the ancient Hebrews and also before there was organized religion to ancient Middle Eastern peoples and to some of the Arabs. Let's take a look at what happened in the land of the Sorbs when the Jews might have lived there and picked up Yiddish from the Sorbs at a time when the Germans were trying to get Sorbs to replace their Western Slavic language with German words, words that might have shown up in Yiddish when the Jews moved to more Eastern towns in Poland.

One of Charlemagne's sons defeated the Wends and burned Bautzen in 806. Wends/Sorbs were pagan, and the Germans were moving in their territory to Christianize them. Also the area geographically had fertile farmland, forests, sand, hills and....potential brides. In the West, the Vikings were on the move also looking for trade and new lands for farming. They sailed their long boats down the rivers to Kiev in the East, eventually to Iceland in the West. What migrations did the Jewish communities at the time begin? You have Jewish scholars and traders from Spain visiting Russia, Syrians in the land of the Norsemen as ambassadors, and Ashkenazi scholar rabbis in France and Germany speaking old French, old German, and writing in Hebrew (or Arabic in Spain and the Middle East, Judeo-Persian in the Caucasus, Persia and Central Asia).

That's the same time that Khazaria reached its golden age and the nobles converted to Judaism much further East where the Volga meets the Caspian in Khazaria. They were trading with Kiev in the Ukraine by then and had a vast empire stretching from the Black Sea across the Caucasus to the Caspian. They inhabited parts of Georgia and Armenia, trading with the medieval Jewish population in Armenia. At the other end of Eastern Europe, here you have the defeated Sorbs. Charlemagne was Christian. The defeated Sorbs/Wends might have converted to Judaism in the 9th century just as the Khazars converted to Judaism at least in part (their nobles and some of the Slavs ruled by Khazars) around that time also. Jews from Byzantium and Persia, Anatolia, Armenia, and Baghdad moved into Khazaria.

Then as the Ashkenazic communities in the Rhineland were blooming, by the year 1100 the Wends had been subjugated. Did some of them convert to Judaism because the German Christians subjected them in the 12th century? German nobles dominated the Wendish peasants and relegated the urban Wends to homes outside the walls or to restricted sections of the city. They could become active in society only through German institutions and the German language. The guilds were German. Mercantile activity was conducted in the German manner. Under pressure, especially in the part of Lusatia under Prussian control, many Wends adopted German names and relinquished their Slavic traditions. Only a funny thing happened.

A lot of these German names taken by Sorbs/Wends (Wend is a German word for Slav) sound Yiddish….such as Lavine/Levine. Does Levine come from the word for Levite? (Levi, Lion, Loew, Loeb, Lewin, Lewinsky, Levin, Levinson, Levenstein)? You find the name "Levine" showing up in Austria in a university in 1598. It's listed in genealogy records as being a Christian name with a Prussian origin in 1635. Look up Catholic baptismal records for the name "Levine" in Prussia in the 17th century. It abounds. Why did so many Jews of Bialystock, Poland and other areas in the Pale of Settlement of Eastern Europe or in Germany choose the name Levine? Were they Levites? Why did they pick this Christian, Prussian name? Or is it merely a way of "Germanizing" Levi, according to rules in the 18th century when Jews were ordered to purchase or assume German or Slavic names in their respective countries? What happened to their secret Hebrew names? Did they remain only in the synagogue?

The Christianization of the Wends began prior to the German conquest, but it was vigorously promoted by the Germans. They also followed the Germans in the Reformation. The majority of Sorbs/Wends converted to Lutheranism in 1530 after the Council of Augsburg.

In medieval times, the pagan Sorbs/Wends did not yet have a written language. If they contributed to Yiddish as the linguists propose, the Jews who began to speak Yiddish, whether new Jews or ancient ones, used the Hebrew alphabet. In fact it wasn't until the time of Martin Luther in the 16th century that Luther emphasized using the vernacular.

The reformation and Luther encouraged the Wends to devise a written language, and in 1574 Luther's Small Catechism became the first work to be published in it. Perhaps those Wends/Sorbs/Lusatians who became Jews at a time in the medieval past before they became Christian and before they had a written language, were so happy to assume the Hebrew alphabet and to speak Yiddish, related to the medieval Sorb language in many ways.

What they have in common also is that the Germans required the Sorbs to replace what once were their Slavic words with German words—the Germanic of medieval times. Also, Yiddish has many medieval Germanic words, some Slavic ones, and Hebrew. It's almost as if the Slavic words in Yiddish had been replaced over time with German words.

That's what happened also to the Sorb language over time as Germanization of this Western Slavic community replaced some of the words. Linguists suggest that the Sorb language had some influence on Yiddish that's attributed to 13th century German. Yet the Slavic words in Yiddish show an eastern influence. This now becomes a melting pot, a salad bowl of language so to speak. You get some words from Hebrew, some from Slavic, some from German, and one or two from ancient Egyptian and perhaps from Turkic, such as Khazar 'yarmak' coins.

There is perhaps a Khazari word for prayer similar in sound to "davenin"...for praying..Then there's the yarmulke....Or perhaps yarmulke, meaning skullcap being spoken in the vernacular of Yiddish referring to the Hebrew word "kippot" for skullcap comes from the Ukrainian feast holiday called Yarmaka—or from other word that only historical linguistic research will uncover.

Without DNA evidence of Khazars compared to ancient Israelites and to modern Ashkenazim compared to Slavs compared to Germanic peoples to see who carries whose genes in this mosaic of communities and migrations, no one can make a statement saying a whole community derives from another community because the details of history are carried in the DNA—the mtDNA and the Y-chromosomes, the tissue typing tests, and the tests of racial percentage. So you assume a whole group of people are descendants of another whole group of people—unless you can show it by intensive DNA analysis of the entire genome.

What you can say is that either the Y chromosomes of the majority of Ashkenazi males cluster with Sephardic and Middle Eastern Jewish males or they don't. And you have read the results of ongoing DNA research. And with the females, all you can do is see how the mtDNA fits into a pattern to shape migrations at least until the entire genome is known. You can trace female ancestry of Ashkenazi women from Eastern Europe until you arrive at the single female ancestor of each woman who lived 20,000 or 40,000 years ago in similar Ice Age refugiums in the Pyranees and/or in the Ukraine.

The clan mothers you find gave rise to almost all other Europeans. And if you reach out to that mother's ancestor, she will be found in Western Asia or in the Middle East or Africa before she reached Europe, except for V mtDNA haplogroup who possibly arose 13,000 years ago in the Pyrenees in Europe when her haplogroup split off from H mtDNA. After the Ice Age mtDNA V migrated north to Scandinavia and also is found among the Saami peoples, and a branch went south to N. Africa, found currently among the Berbers.

So you have back migration from Europe to N. Africa. Let's consider the male Levite Ashkenazim of Eastern Europe. Could they be descended partly from Sorbs? Is it this West Slavic people living in East Germany and Western Poland who contributed to the Ashkenazi gene pool and Yiddish language? Would the admixture in Ashkenazim, if any, come perhaps from Sorbs rather than Turkic Khazars? Would Yiddish have more Sorb words than Khazari Turkic words? Would some Ashkenazim even resemble Sorbs more than other Europeans? Or would they resemble more their more ancient Eastern Mediterranean ancestors?

So far no scientist has compared ancient Israelites to modern Ashkenazim and found an unbroken chain of ancestry going back until they found the Cohen Modal Haplotype that does go back—in the males to a common ancestor in the Middle East thousands of years ago, perhaps 1,500-3,000 years ago—about the time of the exodus. And that ancestor had a common ancestor in the Middle East about 7,800 years ago from whom most people in the Middle East are descended.

Could Ashkenazi Levite males be descendants of the Sorbs/Wends/Lusatians? If the genetic analysis is right and they are close genetically to the Sorbs of today, it would show how enriched the tapestry is of the mosaic of various Jewish communities of the world all acting as a catalyst, coming together, intermarrying, and forming a quilt. Let's look at the Sorbian language.

There are two versions of Sorbian, also called Sorbic, Wendish or Lusatian, corresponding to the divisions of the Lusatian region. Both versions belong to the Western Slavic group. The southern area called Upper Lusatia speaks a dialect nearer to Czech (Luther's Catechism was translated into Upper Sorbian); the northern area, or Lower Lusatia, a dialect nearer to Polish.

Which version is closest to Yiddish? Traditionally the Wends call themselves Srbi (Sorbee) in their own language, but the Germans call them Wenden, a term widely used both by others and by, many of the Slavic Lusatians themselves, including those who migrated to foreign lands in the 19th century.

When the Jews left the Rhineland in medieval times to move a little to the East, say in East Germany or Western Poland, Wend was the German name for all West Slavs. The German word, "Wend" eventually symbolized the feared Germanization of the Wends that began in the 9th century with the Carolingians and continued through the Weimar Republic and the Nazi period. So if you were a Srbi a Sorb (Wend) in the 9th century, and a Jewish village from the Rhineland sprung up near you, and you were resisting Germanization, forced Christianity on your Pagan beliefs, wouldn't you at least consider becoming Jewish? You'd automatically get the chance to write your language with newly learned Hebrew letters.

The alternative would be the German alphabet, but you've just been defeated by the Germans and told to become Christian, to replace your Slavic language with German words. What would you do? Comprise, and start speaking what would develop later into Yiddish? Does this possibility sound more plausible than 500,000 Khazars turning Jewish and rushing into Poland after the destruction of Khazaria in 965 CE? Make up your own mind. Chances are a little bit of each might have happened, with traces of this enriched mosaic of genes joining the towns of the Jews fleeing oppression from the Middle East or Spain.

Lusatia was ruled at times by non-German princes. Since the Peace of Prague in 1635, it has remained under German control. In the 19th century when Jews were seeking surnames, many of these surnames came from Prussia such as Lewin, Lewinsky, Levine, Levene, Levin, and Levenstein. Any book on genealogy will give Prussia as the place of origin for many names now used by Ashkenazim. Some genealogy books will give surnames as originated in both Prussia and Austria for these same names.

Before German unification in 1871 Lusatia, land of the Sorbs/Wends was contained in parts of the kingdoms of Prussia and Saxony. Lower Lusatia was under Prussian administration and Upper Lusatia under that of Saxony.

After World War II, the original land of the Sorbs was in part of the German Democratic Republic (East Germany). The Sorb area was divided administratively between the districts of Dresden and Cottbus. Wendish ethnic awareness has been encouraged under the German Democratic Republic. The Slavs of Lusatia, in their own language probably would prefer to be called Sorbs. So the term Sorb has been adopted for the Slavs of Lusatia.

Sorbs may want to reflect their Slavic heritage as West Slavs of Central European ancestry. They are not the Serbs of southern Europe who are another Slavic people living in Serbia. And Sorbs have never lived in Serbia, which is near Macedonia. Sorbs live in Eastern Germany and Western Poland. In current times, nearly 60,000 people in Lusatia call themselves Sorbs. For more information, contact the Sorbian cultural center (the Domowina). Sorbs have a Sorbian-language newspaper, radio station, theater, and folklore tradition.

So what if you are an Ashkenazic Levite whose Y chromosome reveals your ancestry is close to the modern Sorbs? Do you invite a Sorb to your home to share family photos? Do you find out genetically whether the rest of your genes match by markers? Or do you take a DNA test to find out your most recent common ancestor or any medieval ancestor linking you to that Sorb you've invited for dinner exchange of family photos? How important is your DNA test result in your core identity and in your life? There are a lot of travels to ponder.

That's why the historic origin of the Ashkenazic communities is debatable. You have to let the genes tell what they can. You also have to look at linkage disequilibrium (LD). Kruglyak (1999) reported that populations don't keep their LD over long genetic distances unless there have been sharp founder effects called bottlenecks.

Posters sometimes go up at genetics conferences illustrating the bottleneck in Ashkenazic mtDNA—female lineages. And the bottlenecks usually are in the recent past. (Reich et al. (2001).

You look at the last Ice Age (LGM) or you go further back to the out-of Africa migration into the Middle East or West Asia to see bottlenecks in huge populations. The bottlenecks can be specific to male or female lineages. You see them in the mtDNA in matrilineal lineages or the Y-chromosome lineages in males.

If the bottle neck or founder effect is harsh enough, the extinctions severe enough, you see a change in the genes (Wright et al 1999). One example of a great reduction in mtDNA diversity is seen in Georgian Jewish females. When you get severe mtDNA reduction in a population, you also get long-range linkage disequilibrium. Only Georgian Jewish females showing the great reduction in mtDNA are not Ashkenazic females from Central and Eastern Europe, especially North Eastern Europe. Also, South Eastern non-Jews are genetically close to North Eastern non-Jews.

So you have to look at Ashkenazic women and men in their own community and see how close or not close their genes are to the host populations. You already know the males cluster with other Jews and with males in the Middle East in 70 percent to 80 percent of individuals studied. What about the origin of Jewish mothers from Ashkenazic communities? Are they related to anyone in the Middle East? Are they related to one another?

Or are they formed from independent local women in independent communities that eventually merged in medieval times? You be the judge after reading the latest genetic findings. If you want to dig further, read the latest articles. Many findings on Ashkenazic mtDNA are unpublished at the time this book is being written, so there's no way to read the articles of the very latest findings. One thing is certain:

The mosaic of communities forms a rich tapestry of having merged together. By the time of publication, the articles will probably be available to the public in the various journals of human genetics. Read them. I've learned a lot from reading them, and I highly recommend the articles and the books.

For now, until any more information comes in, the mothers of the Ashkenazic communities present a mosaic of communities that have merged. As Dr. Mark Thomas wrote, "Ashkenazi Jews were made up of a mosaic of different, independently founded communities that has since homogenised."

Also, I highly recommend the audio tapes of the conferences on Jewish genealogy. Visit audioiotapes.com on the Web at http://www.audiotapes.com/conf.asp?ProductCon=77 to see the 111 available audio tapes for the 19th Annual Conference on Jewish Genealogy recorded in 1999.

Check out the Y-chromosome haplogroups at: http://freepages.genealogy.rootsweb.com/~dgarvey/DNA/markers.htm

Check out your mtDNA at:
http://www.stats.ox.ac.uk/~macaulay/founder2000/tableA.html
http://shelob.bioanth.cam.ac.uk/mtDNA/toc.html

Also see the geographical index at: http://shelob.bioanth.cam.ac.uk/mtDNA/geogind.html

You can view the CRS online at: http://www.bioanth.cam.ac.uk/mtDNA/anderst.html. Also see the Web site on how to read female-line DNA tables at: http://family.nf/english/projects/female-lines/how-to-read-female-line-dna-tables.html.

Below is the mtDNA of one person descending from a family of orthodox Jewish great grandparents who have been in rabbinical, scholarly families for dozens of generations according to family traditions. The individual's mtDNA mutations below from the low-resolution HVR1 and the high-resolution HVR2 mutations are compared with the Cambridge Reference Sequence (CRS) to check for variability.

HVR1 Haplogroup H
HVR1 Mutations:
16189C
16356C
16362C
16519C
HVR2 Mutations:
263G
309.1C
315.1C
523-
524-

Where is the individual's ancestral geographic origin? Since genetics has both a cultural component as well as a biological, where do these genes appear geographically in the ancient past, the recent past, and the present? Does the individual descend from a matrilineal line in the Middle East, in West Asia, in Northern Europe, or in Southern Europe?

The following is the mtDNA test result from the individual's personal DNA test with ancestral mutations that evolved over thousands of years as compared to the Cambridge Reference Sequence (CRS) shown below (with permission). The HVS-1 and HVS-2 mtDNA test was done by Family Tree Genetics.

Let's take a look at how online research can give some clues, but not a definite decision on where the matrilineal line originated. First we'll look at HVS-1, the low-resolution haplogroup mutations in the mtDNA, that is the matrilineal lines that go back about 21,000 years in Europe, and perhaps are even older before the ancestor migrated to Europe. Somewhere mtDNA haplogroup H hand a previous ancestor, possibly in the Middle East who was pre-HV. That ancestor gave rise to another ancestor HV, which in time gave rise to H.

Finally, H migrated from the Middle East or Western Asia to Europe and settled in the Ice Age "refugium" in southern France or Spain, near the Mediterranean in the foothills of the Pyrenees. Then about 13,000 years ago, V mtDNA (split off from H and migrated to Scandinavia and to the Berbers in N. Africa. H mtDNA became a "clan mother" during the Ice Age by successfully reproducing enough daughters to pass on the haplogroup. After the Ice Age, H expanded all over Europe from Iceland to the Urals and is a "pan European" type today. H mtDNA also is found in the Middle East. About 46% of Europeans and about 25% of Middle Eastern peoples currently have H mtDNA. V mtDNA possibly arose in Europe.

Searching the Mitomap Table at: http://shelob.bioanth.cam.ac.uk/mtDNA/ hvr1n.html we find the slowest mutating sequence, 356 shows up in individuals from the following geographic locations: UK, Havik, Karelina, Portuguese, Turkish and Volga-Finnic.

Then we look at the next sequence which would be 356C and 362C and find it shows up in a Bulgar and a Karelian. When I get to the exact sequences 189C, 356C, 362C, it shows up in a person from the UK, Spain, Austria, and Bulgaria.

We see that the main mutation is 356 or actually 16356C in the individual with mtDNA H haplogroup, mutations at 16189C, 16356C, 16362C, 16519C. The 189 mutation or 16189C is a fast mutating site. So is the mutation 362 or to be exact, 16362C as it's usually written. So what we focus on for ethnicity is the slowest mutating site, which is 356 or 16356C.

In the tables at http://shelob.bioanth.cam.ac.uk/mtDNA/hvr1n.html on the Web, we see that 16356C is found in the Volga Finnic people of North Eastern Europe, but also in Turks, Bulgarians, and Portuguese. So does this very frum (religiously obedient) orthodox Ashkenazi Jewish person with ancestors from Bialystock, Poland and Bessarabia Rumania have a female ancestor who in ancient times—very ancient times came from Karelia (nation within and just east of Finland)? Or would the individual's Yiddish speaking ancestors have come from Bulgaria, Turkey, or Portugal? People of Southeastern Europe are closely related genetically to people of Northeastern Europe.

The time frame here could be tens of thousands of years or as recent as a common ancestor perhaps 250 years ago. Let's look a bit deeper into the ancestry and the sequences. Keep in mind that H haplogroup is pan-European and found all over the Middle East. The sequences place the individual in various possible geographic locations as far as ancestry.

These particular sequences also show up in one to four individuals recorded in other databases in Siena, Italy, in Crete, in Norway, Germany, Poland, Scotland, Bulgaria, Spain, England, Iceland, Croatia, Albania, Portugal, Austria, Bashkortostan in the Urals, the Ukraine, and among the Komi of Finland.

How can you tell where (geographically) the person's female ancestral line originated? That's part of the history and geography of human genes.

You can see where the mtDNA haplogroup started out, going from the Middle East to Europe, probably to the north of Spain or south of France 20,000-21,000 years ago. You see H in the Middle East about 28,000 years ago in the Fertile Crescent and in Anatolia and what was Armenia. You see it in the refugium in the Pyrenees 21,000 years ago, and in England 12,000 years ago where a young male skeleton was found in Gough's Cave in Somerset having H mtDNA haplogroup.

At the end of the Ice Age 12,500 years ago, H mtDNA traveled from the Cro-magnon caves of southern France and Spain to what was to become England. All people in with European ancestors with H mtDNA descend from one single female with H mtDNA haplogroup who lived about 20,000 years ago in the Pyrenees area where there was a "refugium" from the Ice Age where the fishing was abundant and the campsites offered good shelter.

Yet, we still have to look further to get a handle on what happened to these particular sequences of H mtDNA. We know by looking at databases of Eastern Europe that those particular sequences 189C, 356C, 362C, 519C are found in

North Central and North Eastern Europe primarily. They show up in Europe at a very low level, but where they show up most frequently is in Germany, Poland, Scotland, Norway, and Sweden.

Today, matches for these sequences are found frequently in England, Belgium, the Netherlands, the US, and Germany. The sequences are found all around the Adriatic Sea. So it's pan-European. On a low level, the sequences also show up in Turkey, Bulgaria, England, Spain, Portugal, Crete, and Italy. On a slightly higher level the sequences show up in Germany, Scotland, Poland, the Ukraine, and among the Komi in Finland.

We know Finland has one of the least diverse mtDNA pool. So all we can surmise from looking at the mtDNA databases is that the mtDNA haplogroup with those particular sequences show up at a low level in Europe and show up at a slightly higher level in North Central and North Eastern Europe. The difference in levels would be four individuals from Scotland and Germany have this sequence verses three individuals from Poland, Norway, Sweden for example. And the source would be a database with about 10,000 sequences in it.

We also know that without examining any particular sequence, H mtDNA in general, appears 25% of the time in the Middle East and more than 46% of the time all over Europe—from Iceland to the Urals. Those particular sequences of the individual do appear in Iceland, Norway, Sweden, Scotland, England, Germany, the Ukraine, (Komi group) of Finland, and Austria….but also in Bulgaria, Turkey, and as far east as Bashkortostan in the Urals. So we have to look at where it appears most—North Central Europe and North Eastern Europe.

The problem in looking at ancestry is that the point of origin is not necessarily where the mtDNA sequences appear in the largest numbers today, but where it is found most diverse. The more diverse the sequences the longer it has been in that place. This gives a clue to the origin or coalescence point.

H mtDNA in Turkey (ancient Anatolia) has a coalescence point of about 28,000 years or more compared to 20,000 years in Europe. However, the sequences of H in the Middle East would be different from the sequences of H in Europe in many cases, but not all.

You have this little movement call back-migration, where from ancient times people traveled from Europe back to the Middle East and stayed there. And you have waves of migration moving from southwest to northeast. About 15% of the Middle East is comprised of Europeans moving back there in prehistoric times, and about 20% to 26% of Middle Easterners, mostly Neolithic farmers after the end of the Ice Age moved into Europe. It took them much longer to reach the northern countries, but some did arrive and stay, introducing by trade the idea of farming to the Paleolithic hunters.

Where else can you check the history and geography of your mtDNA or Y chromosome? There are several excellent databases on the Web where you can check your sequences to see how many individuals from which countries have the same sequences as your own. Keep in mind that this information does not necessarily tell you where your particular ancestors came from, but the databases give you an idea where sequences like yours are found.

The question is, when you find someone with your sequences, does it mean that at one time you and that person shared the same common ancestor perhaps 300 generations ago or more? If you share the same single female mtDNA H haplogroup ancestor who lived 20-21,000 year ago in Europe, yes, you share at least the same first H mtDNA clan mother.

And perhaps if your sequences match today, you share a slightly more recent ancestor. However, those mutations could pop up anywhere in anyone's line, related or not with or without the same mtDNA H. It all depends on what the other mutations are along with your sequences. So you'd have to check out that person's lineage to see how it might have crossed the path of your ancestors to get a handle on your search.

Let's look further at HVS-1 sequences. First I searched the Internet at Macaulay's Founder Tables to see where my exact mtDNA sequences are showing up at: http://www.stats.ox.ac.uk/~macaulay/founder2000/tableA.html

Be aware that this table from the United Kingdom may not necessarily contain the same sequences from the same countries as some of the databases in Eastern Europe. So check out as many databases online or in scientific journals as you are able to access. For example, I found the exact sequences in Macaulay's Table listing the following sequences found in individuals claiming ancestry from the following geographic areas 189 356 and 362 for England, Spain, Austria, and Bulgaria. The tables at the Web site list abbreviated names for the countries.

How you search in what country currently your mtDNA HVS-1 haplogroup appears is by realizing that the letters stand for names of countries. For example:

Nu means Nubia. Eg stands for Egypt. Be stands for Bedouin. Ye is Yemen. Iq is Iraq. In is India. Sy stands for Syria. Pl stands for Palestine (Arab Palestinian). Dz is Druze, a tribe of people living in Lebanon. Tk is Turkey. Ku is Kurd. Ar is Armenia. Az is Azerbaijan (in Central Asia), Cauc stands for people living in the areas of the Caucasus mountains such as Georgia, Armenia, Ossetia, Chechnya,

Kabarda, Adyge, the Circassians, for example, the peoples of the north and south Caucasus mountains. MdE stands for Eastern Mediterranean.

SE is Southeastern Europe. MdC is central Mediterranean—Italy. Alp means the area of the Alps, such as Austria. NC stands for North Central Europe—Germany. MdW stands for Western Mediterranean, such as Spain and Portugal. Bs stands for Basque. NW stands for Northwestern Europe, namely, England/Britain. Sca means Scandinavia—Sweden, Iceland, Norway, Finland, and Denmark. NE is North Eastern Europe.

Look at the sequences of H mtDNA 189, 356, 362. The mutation (519C) wasn't included as some mtDNA research testing only test 400 or so base pairs. You'll see where the exact sequences appear in the table in only four countries: Spain, England, Austria, and Bulgaria. If you look at the sequences that contain the 356C mutation without the 362 mutation, you'll see it appears in more places and that more people have that sequence. The conclusion is that H mtDNA 189C, 356C, 362C, 519C appears at low levels in Europe. Perhaps it went through a bottleneck. Perhaps it went through extinctions, even though it's the most frequent mtDNA haplogroup in Europe—which is H.

However, it does appear in Europe in those four areas in that particular database. In an Estonian unpublished database, it appears in many more countries, including the one's in Macaulay's database. The Estonian database has those exact sequences appearing among the Komi of Finland, in the Ukraine, in Croatia, Albania, Bulgaria, Turkey, Crete, but more frequently in Germany, Poland and Scotland, and also in England, Iceland, Sweden, Norway, and in Bashkortostan, a country in the Urals.

According to the Macaulay database, those exact sequences aren't found east of Bulgaria. In the unpublished Estonian database Turkey and Bashkortostan are added, but also Crete. Since Albania and Crete show these sequences in the Estonian unpublished database, history notes there was a migration in medieval times from Albania to Greece and to Crete, and the population was absorbed.

So what are the origins of this person—Volga Finnic? Germanic? Polish? Bulgarian? Can we ever know for sure? The exact sequences are found in England, Netherlands, and Belgium. Where did the most recent ancestor of this individual originate? Will we know more in the future about tracing our ancestors from a great grandparent back in time several hundred or several thousand years?

We have the first ancestor—H mtDNA and the present, with the particular sequences living in the US. We know for generations the ancestors lived in Poland and then in Rumania. When did the ancestors arrive in Poland, and did they come from Germany or Lithuania? If Lithuania, when did they arrive there from what other geographic place? It's a fascinating migration as the sequences reveal.

So if you want to match your own sequences to a table that shows what nations your sequences appear in today, click on Macaulay's statistical tables from Oxford University in England and see whether your particular sequences are found in that database. The Web site again, is: http://www.stats.ox.ac.uk/~macaulay/founder2000/tableA.html.

Note that to the left of the table all the countries of the Middle East are listed starting with Nubia (in Africa) at the extreme left with Egypt next and then Bedouins, Yemen, Iraq, India, Syria, Palestine, Druze, Turkish, Kurdish, Armenian, Azerbaijan, Caucasus, Mediterranean East. Then you have to the right half of the table all the European countries going from the central Mediterranean—Italy north to the Alps and Austria followed by Germany, then down to Spain and Portugual and then to the Basques in Western Europe, then up again to England in NW Europe, then northeast to Scandinavia and finally to North East Europe.

The table more or less follows several of the routes of migrations of ancient and prehistoric peoples out of Africa (Nubia) into Egypt an then into the Middle East, to India, and migrating west to the Middle East and then north to the Caucasus, then back to Europe and finally, the last migration route from southwest Europe to Northeastern Europe passing through central Europe, England, Scandinavia, and finally reaching Northeastern Europe. Where are you on any of the databases? Also check out the mtDNA Concordance Tables at: http://shelob.bioanth.cam.ac.uk/mtDNA/hvr1o.html.

The individual from a long line of orthodox Jewish families was concerned that the mtDNA sequences didn't show up on the left side of the Macaulay Database statistical tables. Why didn't they show up with an ancestry in Iraq, Palestine, or Syria? Did the Jews flee oppression from those areas and move into Western and then Eastern Europe?

Why is the ancestral mother matching with all those Central and Eastern Europeans and also with individuals in Spain and Portugal and England? Note that only one person from Spain, one person from Austria, one person from England, and one person from Bulgaria shows up for that exact mtDNA sequences of HVS-1. So we dig a little further and look at the high resolution mtDNA mutations for HVS-2. And what do we find? This table below is from the individual's personal DNA reading for the high-resolution HVR2 mutations regarding deep maternal ancestry.

HVR2 Mutations   263G
                  309.1C
                  315.1C
                  523-
                  524-

We find that the HVR2 Mutations (or HVS-2) are found in most people all over Europe and all over Western Asia and the Middle East, or are they? Look where they do appear. The HVR2 mutations also are very common sequences. Remember that the HVS-1 appears only at a low level in Europe. But look at the high-resolution HVR2 mutations. They appear in France and in the Orkney Islands, in Scotland, Bulgaria, and Turkey.

HVR2 can distinguish mtDNA haplogroup H from mtDNA haplogroup U4. If you see the letter "A" at position 00073 in HVR2 (or HVS2) the mtDNA haplogroup is H. If you see a "G" at position 00073 in HVR2, the mtDNA haplogroup is U4. Both H and U4 contain the mutation sequence 356 or 16356C. However, U4 and H show up in different countries but they also show up in the same geographic ancestral areas. For example, U4 doesn't show up among the Ossetians of the Caucasus Mountains, but H shows up among this ethnic group. Sequences 189, 356, 362 haplogroup H doesn't show up in the Caucasus, at least in the databases, but again, you have to think about how many people are in the sample. Some databases contain at least 10,000 individuals from a variety of countries.

Check out some of the founders at: http://www.stats.ox.ac.uk/~macaulay/founder2000/tableB.html.

Note that the tables of founders at V. Macaulay's site of Oxford University, UK is excellent for checking your sequences on the Internet. I highly recommend looking at all the tables on these sites. So, is this individual with mtDNA haplogroup H, sequences at 16189C, 16356C, 16362C, 16519C Germanic, Northeastern European, Southeastern European, Middle Eastern or what do you think? What we are looking at here is the person's ancient female ancestor with the latest mutations since H mtDNA evolved thousands of years ago and is found all over West Eurasia.

What we have to narrow down is the current mutation sequences. That's the mutations or transitions in the sequences that evolved away from the Cambridge Reference Sequence (CRS) and see where in geographic location the sequences as they appear today turn up. Most helpful are tables online you can check out and the various statistical databases that match your mtDNA sequences to the actual parts of the world in which they appear. However, it won't tell you for sure whether you're related to anyone in that location today. Check out Vincent Macaulay's Web site at:

http://www.stats.ox.ac.uk/~macaulay/ and look over his tables. Perhaps you'll find your mtDNA sequences there. Also for males, interested also in Y chromosome haplogroups, check out the various Y chromosome tables at: http:// freepages.genealogy.rootsweb.com/~dgarvey/DNA/markers.htm

If you want to see how genetically close Ashkenazim are to other nationalities, look at the Human Races Calculator on the Web at:

http://racearchives.com/calc/jewsychroms.asp?popid1=1&popid2=9&dbname= jewsychroms

Compare 6 Jewish and 23 non Jewish Y-Chromosome Analysis. You'll see that Ashkenazic male Y chromosomes match 95% with Syrian male Y-chromosomes, 94% with Greek Y-chromosomes, and 95% with Roman Jews.

Also see at that site:

http://racearchives.com/calc/haplo_profiles.asp?dbname= europeanmtdnabyregion&po

You can select a population and view its profile as far as how distant or close the population is from another population. You can also view Y-chromosome haplotype frequency tables from the Hammer study at http://www.rasch.org/ rmt/rmt161g.htm. The article is titled, "DNA and the Origins of Jewish Ethnic Groups. In the article at this Web site, the Ashkenazi male Y chromosomes are positioned on a table between the Jewish non-Ashkenazi and the Turks. Other studies report Ashkenazi male Y chromosomes close to the Greeks and still other studies close to the Kurds. So you have to look at a large number of studies to see whether there is agreement over time where the Ashkenazi stand.

The article, "In DNA, New Clues to Jewish Roots," by Nicholas Wade, at: http://www.racesci.org/in_media/raceanddna/dna_jewish_nyt_May2002.htm reports that the Ashkenazim show "less diversity in its mitochondrial DNA, perhaps reflecting the maternal definition of Jewishness. But unlike the other Jewish populations, it does not sow signs of having had very few female founders."

What the pattern in the Ashkenazic population does suggest is that it's a mosaic of separate populations "formed the same way as the others." According to the article, it reports that Dr. Harry Ostrer, a medical geneticist at New York University said, "the 26 specific genetic diseases found among Ashkenazim usually attributed to 'founder effects' could be explained by the idea of small populations."

So what you get from reading more and more research articles on Ashkenazic genetic origins is that the whole idea of Jewish communities founded by Jewish men and local women is against the Jewish tradition. What do the genes say? Tradition says the first Jewish communities in Eastern Europe were formed by families who fled oppression from one land and were invited by princes to populate other lands, usually to the East.

What was the primary catalyst that brought the Jewish communities together in medieval times through the 18th century? How do you trace Jewish DNA for ancestry and family history through 3,000 years by looking at a merging mosaic of communities? How much trade, traffic, intermarriage, migration, and rapport occurred between the Ashkenazic and non-Ashkenazic Jewish communities of Europe, the Middle East, Africa, Central Asia, India, and China in historic times? And when the records can't be readily accessible, how much detail can we read in the DNA?

CHAPTER 8

▼

MERGING A MOSAIC OF COMMUNITIES BY DNA TESTING

Say some people you know are or you are Jewish with ancestors from Eastern Europe—Ashkenazim whose ancestors spoke Yiddish in the old country. How do you interpret your own DNA test results to look at not only your own family history, but your ancestors as far back as you can go genetically? How do you look at your ethnicity in the light of choosing what foods to eat with two separate types of DNA tests—one for ancestry or origins and one for nutritional genomics—finding risks and preventing or delaying problems through nutrition? It's a consumer's right to explore all avenues of your genetic signature with DNA tests for ancestry and tests of various SNPs or your entire genome for finding out risk and tailoring what you eat to how your genes respond.

You know the records may stop at one point back in time, but the DNA will not stop. It will mutate over thousands of years, but it will show you an open road on where you came from, even in a general sense.

How can you apply what you learn about your genes to improving your life, your nutrition, and your knowledge of your origins? Gene tests will not tell you what religion you are or what your relatives converted to a thousand years ago, but you'll learn far more important information for your time capsule, how to customize what you eat to the entire history of your ancestry, your individual genetic signature or tapestry. You'll need two kinds of tests—a nutritional genomics test and a DNA test for ancestry.

I highly recommend the book, *The Beginnings of Jewishness* by Dr. Shaye Choen, professor of Jewish literature and philosophy at Harvard. When you begin to trace your DNA think about when the Jewish communities were formed. Were they formed in Southern Italy in the third and fourth century? Or were the communities formed soon after the Islamic conquests from the seventh to the ninth centuries CE?

If you consider Jewish tradition that says families practiced matrilineal descent to determine Jewishness, and your DNA search takes you to the seventh to ninth centuries CE, you can see how small communities were formed in places such as Alsace in France or in the Rhine Valley. Look at articles written by Dr. Hammer, Dr. Goldstein, and the population geneticists who are researching Ashkenazic mtDNA, the founding mothers.

Look at DNA to see founding events. Read the article by M.F. Hammer, A.J. Redd, et al. (2000) "Jewish and Middle Eastern non-Jewish populations Share a Common Pool of Y-Chromosome Biallelic Haplotypes." Proc. National Academy of Science, USA, Vol 97, Issue 12, 6769-6774, June 6, 2000. Then compare the study with later articles such as ""High-resolution Y chromosome haplotypes of Israeli and Palestinian Arabs reveal geographic substructure and substantial overlap with haplotypes of Jews." Oppenheim, Ariella, et al. *Human Genetics* 107(6) (December 2000): 630-641.

Check out "The Y Chromosome Pool of Jews as Part of the Genetic Landscape of the Middle East." Oppenheim, Ariella, et al. *The American Journal of Human Genetics* 69:5 (November 2001): 1095-1112. I also recommend reading the following articles: "Study finds close genetic connection between Jews, Kurds." Traubman, Tamara. *Ha'aretz* (November 21, 2001).

According to the article, it's the Sephardic Jewish males who closely resemble the Kurdish Jewish males. The Ashkenazim were found to differ from the Sephardic and Kurdish males, even though in most previous studies, the Y chromosome of Ashkenazim clusters closely with Sephardic Jews. Yet in this study the Kurdish and Sephardic Jewish populations differed to some degree from Ashkenazic Jews. Studies are still suggesting Ashkenazic males mixed with European peoples during their diaspora in varying degrees, except for the Ashkenazic males with the Cohen Modal Haplotype on their Y chromosomes.

The researchers suggested that the approximately 12.7 percent of Ashkenazic Jews carrying the Eu 19 chromosomes—signature of Eastern Europeans, descend paternally from eastern Europeans. The Eu 19 chromosome shows up in 56% of Polish men and 60% of Hungarian men. Not all Ashkenazic males carry the Eu19 chromosome.

Read the article, "Are today's Jewish priests descended from the old ones?" Zoossmann-Diskin, Avshalom. *HOMO: Journal of Comparative Human Biology* 51:2-3 (2000): 156-162. See the article at: http://www.ariga.com/genes.shtml titled, "Doctor finds fault in the contentions that the "Cohen modal haplotype" designates Israelites and that most Jewish priests have a common ancestor." 27 February 2001.

According to the article at the Web site, "Zoossmann disputes point that the Cohen Modal Haplotype is the signature for the Hebrew nation.*The Cohen modal haplotype is the most common haplotype among Southern and Central Italians*1, Hungarians*2, and Iraqi Kurds*3, and is also found among many Armenians*4 and South African Lembas*5. This calls into question the notion that the haplotype was a marker for the ancient Hebrew population."

You'd have to compare Dr. Zoossmann's article with the articles that discuss the Cohen Modal Haplotype such as "*Y Chromosomes of Jewish Priests,*" in Nature volume 385 by Michael F. Hammer, Karl Skorecki, and their colleagues, January 2, 1997, and the article, "*Origins of Old Testament Priests*" by Karl Skorecki, David Goldstein, et al. in Nature volume 394. Before you begin your research, look at the Cohen Modal Haplotype Web site at: http://www.geocities.com/ hrhdavid/cmhindex.html.

The point is that whenever you have a research study and a resulting article, another scientist will find some flaw in it. That's the way the scientific process moves forward. My highest recommendation on how to understand all this is to look at articles and research done by scientists with enough genetics background to consider the possibilities. There are many sites on the Web discussing Jewish genetics, many written by historians and many written by scientists and physicians. Then there are articles summarizing secondary sources. You have to do your own homework and be the judge. New studies come out each year with new information. The door is open for new results and findings.

The majority of Ashkenazic Jewish men carrying the Eu 9 chromosomes and similar markers descend paternally from Judeans who lived in Israel two thousand years ago. So the point is you can't make a judgment about the ancestry of any particular Jewish male—or female—until you look at all the genetic markers and match them to the geographic location where they are or were.

There's always the Cohen Modal Haplotype found in some Ashkenazic males to research. I recommend the article, "The Priests' Chromosome? DNA analysis supports the biblical story of the Jewish priesthood." Travis, J. *Science News* 154:14 (October 3, 1998): 218. Jewish women don't have a Cohen Modal Haplotype because it's a male Y-chromosome passed from father to son for thousands of years. The new puzzle to solve is the Ashkenazic females. Since Jewish tradition passes the Jewish heritage as a religion and ethnic community from mother to children, what geographic areas did the women come from?

I suggest following studies on mtDNA and other markers to see where the mothers came from or at least who the mothers' genes are closest to in current times. So much has been studied and written about male Y-chromosomes and ancestry and so little on women. The door is just opening and currently much of the data is yet unpublished.

So patience will reveal new information on women as soon as the newer studies are published. In the meantime, keep informed by looking at the latest articles in the various human genetics journals. Contact your local Jewish genealogy society and see what speaker is available who is researching mtDNA and other female ancestry markers. Universities around the world have population genetics departments doing research. Keep in contact with them and their publications and/or conferences.

Once you found your DNA trail—mtDNA for women, Y-chromosomes for men, and even if you took a racial percentages DNA test, what do you do with the information? You create a time capsule, a scrap book, a type of DNA and oral history journal. You use photography, text, sound, music, speech, video, and you put your ancestral record archive in an oral history—transcribed and put to audio or video, to sound, text and imagery.

You even may want to put your information on the Web or in an archive to be viewed by future descendants. The whole tapestry links you back to that first ancestor and to the one in the future. It's a rich experience in history and time.

How do you write, tape, and transcribe an oral history of your DNA along with your genealogy and family history records and photos? Assuming you're a beginner in genealogy with no science background and interested in family history, where do you begin your search? What's the cultural component behind a trait as biological as your genes?

If you're a family historian, an oral history researcher, or a person fascinated with ancestry, here's how to understand the results of DNA tests. Different people have different, sometimes opposite opinions on whether DNA testing is a useful tool in the hands of family historians. If you are a carrier of a genetic disorder, DNA testing is useful in researching your family history to find out who was the first carrier in your ancestry back in time.

Here the debate unfolds as scientists, authors, physicians, media people, owners of DNA testing companies, genealogists, historians and researchers comment, write, and opine on DNA testing and genealogy.

What is the distribution of mtDNA haplogroup K? We know that haplogroup K is a clade of haplogroup U. Previous studies revealed that 27% to 32% of Ashkenazim have haplogroup K and that 9.0 of Ashkenazim have the mtDNA haplogroup H that follows the CRS. The rest have mutations of haplogroup H that does not follow the sequences of the CRS without mutations. What is the distribution of mtDNA haplogroup K in the rest of the populations sampled?

Take for example one sample appearing in the chapter, *Anatolian and Trans-Caucasus Populations*, Archaeogenetics: DNA and the Population Pehistory of Europe, McDonald Institute Monographs, 2002, page 223. The highest figure is 31.7% in Georgians sampled have haplogroup K mtDNA. Next are Armenians sampled with 27.8% haplogroup K mtDNA. Following are Turks with 24.7%, and Siena, Italy samples, Tuscans, Saradinians, French, and Albanian samples with 35.8% haplogroup K mtDNA. Haplogroup K mtDNA also appears at 25% in the Nile Valley, and at 20% in Ethiopians, and 1.2% of Indians. It also appears at 11% in Estonians, Finns, and Karelians, and at 18.6% in Russians, Poles, Czechs, and Slovaks.

Where did it originate? In the Middle East, probably, but about 17,000 years ago it developed in Europe in the area where Venice, Italy is today and then migrated into the Alps.

How do you use DNA testing to interpret family history records? How eager are people to take a DNA test for family history research? Most DNA tests require only that someone swish mouthwash around in his or her mouth and send it for testing to a laboratory. So what can a DNA test really tell you about your own ancestry—distant or not so distant? And most of all, how do you interpret and use the results?

Here's a letter from Dr.Mark Humphrys:

Lecturer
School of Computer Applications,
Dublin City University,
Glasnevin, Dublin 9, Ireland.

"Dear Anne:

Here's a summary of the position as I see it:

Why everybody in the west is descended from Charlemagne:

We all know that all humans are related. So a good question is: When was our Most Recent Common Ancestor (MRCA)?

Surprisingly, the answer to that question is a lot more recent than DNA studies would suggest, since we are searching all lines of descent, rather than just the lines genes traveled on. You do not inherit all of your ancestor's DNA, but only a small part of it. And yet, even if you inherit NONE of their DNA (which is not only possible, but probable, as you go back far enough), they are still your ancestor.

To find the answer to the MRCA, we need to look beyond DNA studies. Mathematical models suggest that, if humans picked mates randomly, the MRCA is in historical times, perhaps c. 1200 AD! This is an amazing result, suggesting that we do not have to go back into prehistory to find an ancestor of every single human! But obviously humans do not pick mates randomly— they tend to mate with people in their local geographic area.

Computer simulations that take this into account suggest that even with a high degree of local mating, the MRCA is still in historical times, perhaps c. 300 AD. If we consider just the West, the MRCA may be as recent as c. 1000 AD.

How realistic are these models?

Well, there has been a growing collection of REAL, proven descents from medieval figures in genealogy. For instance, my own children have proven descents—through many different lines—from Charlemagne, who lived around 800 AD.He is the ancestor of most of the royal houses of Europe and so is a natural focal point for the genealogies of the West.

My web page, Royal Descents of Famous People, is a large and growing list of famous people in the West who are all proven descendants of Charlemagne.And if all these people have proven descents, every step of the way, how many more people must have descents in reality that cannot be proved because of the scarcity of records? It must be a much greater number.My work is a strong indicator that everyone in the West descends from medieval royalty.

In short, work by a number of people—my genealogical study, other people's computer simulations and mathematical models—all confirm each other's findings that the MRCA for a large, interbreeding area such as the West, is within recent recorded history. Another finding of those models is that not long before the MRCA, if someone is the ancestor of anyone alive today, they are the ancestor of all people alive today. Since Charlemagne is probably around that early date for the West, and since he is a proven ancestor of some people, it is likely that he is the ancestor of all people in the West.

Everybody in the West is descended from Charlemagne:

In conclusion, if you have west European ancestry at all, it seems virtually impossible for you not to be descended from Charlemagne, who lived around 800 AD. 90 percent of the world (including all the West) is descended from Confucius:

For the MRCA of the whole world, we need to consider extremely isolate daboriginal populations. If they were truly isolated, we may have to go back thousands of years to get a common ancestor with them. For people who did not live in isolated enclaves though—the West, Middle East, more or less all of Asia, most of Africa—the MRCA is highly likely to be in recent historical times (late BC, possibly even AD). Anyone with ancestry from these areas is, for example, almost certainly a descendant of Confucius,who lived around 500 BC and who is a proven ancestor of some people alive today in China, hence probably ancestor of all people in the world except the extremely isolated.

This exciting consensus is fairly new, and is supported by three independent fieldsof (a) genealogy, (b) mathematical models, and (c) computer simulations.

The findings are robust with respect to barriers such as religion, class difference, etc.All one needs is a tiny amount of crossing of such barriers in the population in the past in order to get everyone today (of different religions etc.) with a recent common ancestor.

The only thing that can push back the MRCA before historical times is total geographic isolation of populations from each other, which we know did not happen for most of the world. People are inspired (rightly so) by DNA studies of ancient common human ancestors tens of thousands of years ago. And yet the fact is that we are almost certainly all descended from any historical figure in classical times that left descendants."

———

end quote

———

Web pages on MRCAs:
http://computing.dcu.ie/~humphrys/FamTree/Royal/ca.html
http://computing.dcu.ie/~humphrys/FamTree/Royal/ca.genetic.html
http://computing.dcu.ie/~humphrys/FamTree/Royal/ca.math.html

http://computing.dcu.ie/~humphrys/FamTree/Royal/famous.descents.html

The Web pages include a fantastic computer simulation by a gentleman named Rohde at MIT to work out the MRCA for a non-random mating model. He confirms much of Chang's work, and in general it is another strong indicator that the MRCA for the world (or 99 percent of it) is post-1000 BC—maybe even AD.

Regards
—Mark
Dr.Mark Humphrys

Lecturer
School of Computer Applications,
Dublin City University,
Glasnevin, Dublin 9, Ireland.
http://computing.dcu.ie/~humphrys/

*   *   *   *

Steve Olson, author of the book, Mapping Human History in a telephone interview with me on December 3, 2002 at 1:50 PM EST answered my question—What do you say about using DNA as a tool for genealogy—to extend family history research?

"The most valuable use of DNA testing is to demonstrate how closely related we are to each other, both as individuals and as members of human groups," Steve Olson says. "I'm a skeptic about the Seven Daughters of Eve (book) because I believe that everyone in the world is descended from all seven of those women. As I point out in my book, Mapping Human History.

"As I point out in the book, I believe that everyone living today is descended from most people who lived a few millennia ago," Olson explains. "So genetic tests need to be interpreted very carefully or you draw false conclusions about being descended from a relatively small number of people."

Does Steve Olson think DNA testing as a tool is useful to genealogists? "No, I don't feel DNA testing can tell you things that can't be discovered in other ways. I should probably say here, though, that I'm fairly skeptical about DNA testing for genealogical purposes, and I'm particularly critical of the Seven Daughters of Eve idea."

The Seven Daughters of Eve, is a book written by Bryan Sykes, from Oxford Ancestors. Bryan Sykes, MA PhD DSc, is Professor of Human Genetics, University of Oxford, and with Oxford Ancestors, who comments in this book. Sykes has a different opinion about DNA testing and genealogy/family history research.

Oxford Ancestors is the world's first company to harness the power and precision of modern DNA-based genetics for use in genealogy. The motto on the Oxford Ancestors Web site at:

<http://www.oxfordancestors.com/> reads: "Putting the genes in genealogy."

Oxford ancestors is based on more than a decade of research into human populations and their origins carried out by Professor Bryan Sykes, Professor of Human Genetics at the University of Oxford and his team in the world-renowned Institute of Molecular Medicine in Oxford, England.

So as you read in this book, and any other books you research, there are many different experts in genetics, medicine, and the media with different opinions.

Science is supposed to be skeptical by nature as well as open-minded for change. Facts provide an open door to further inquiry, and facts need to be checked as new information comes in. So what stance will you take? Will you take the skeptical side on DNA testing? Or do you reason that DNA testing is one more tool that adds genes to genealogy? Where will you stand in this debate—observer, skeptic, or DNA researcher working with genealogy and genetics?

Perhaps you're a genealogist or oral historian doing family history research and you have no science background in genetics. Now that the molecular revolution has introduced DNA testing for ancestry research, here's how to use your curiosity.

You may want to find information about tests of racial percentages or DNA testing and ancestry. Make up your own mind. Now you can combine DNA research and oral history research into one archive. How do you interpret DNA tests and how do you plan, record, and transcribe an oral history for your family or others?

DNA testing for ancestry is offered by companies that combine DNA testing (done by commercial laboratories and/or university laboratories) with surname genealogy.Various surname groups on the Web also offer discounts with DNA testing companies and laboratories. Ask for the facts.

Whom do you go to first, assuming you have no background in either science or genealogy? Scientists? The media? DNA testing companies? University laboratories? Oral historians? Genealogists? Biological anthropologists? DNA mailing lists on the Web? Whom can you trust?

And where does the truth lie—in technology? It's like twelve blind men asked to describe an elephant. Each touches a different part of its body and replies, "it's a tail," or "it's a trunk." Everyone has a different opinion of the value of DNA testing in researching family history or ancestry.

What do the changing scientific facts say, for now? Question authority, and do your research. Then decide what tools are the best for you in your own family history quest.

Go online to Pub Med, an NCBI research engine on the Internet for medical and scientific articles in professional journals. Read the article titled, "The common, Near-Eastern Origin of Ashkenazi and Sephardi Jews Supported by Y-Chromosome Similarity," Santachiara Benerecette, AS, Semino O, Passarino G, Torroni A, Brdicka R, Fellous M., Modiano G, Diparrtimento de Biologia Cellulare, University della Calabria, Cosenza, Italy. Annals of Human Genetics Jan; 57 (Pt 1): 55-64, 1993. Compare the study to studies done on Jewish Y-chromosomes a decade later. What new information did you find?

In the 1993 study about 80 Sephardim, 80 Ashkenazim and 100 Czechosolvaks had a DNA test of their Y-chromosomes. These males were examined for the Y-specific RFLPs. The aim of the study was to look at the origin of the Ashkenazi gene pool by looking at specific genetic markers useful to estimate the paternal gene contribution to these men's origins. To find out ancestry, scientists examine specific genetic markers in the DNA.

The genetic markers of the Ashkenazim and the Sephardim were compared with each other and with the Czechoslovakians. The Czechoslovakians turned out to be very different from the two Jewish groups. Who are the Ashkenazim closest to? The Sephardim first, and then the Lebanese. Both the Ashkenazim and the Sephardim appeared to be closely related to the Lebanese. The conclusion of the study was that Ashkenazi and Sephardi are both close genetically to the Lebanese as far as their Y-chromosome-specific RFLP markers. According to the Pub Med abstract, "A preliminary evaluation suggests that the contribution of foreign males to the Ashkenazi gene pool has been very low (1% or less per generation.)

Also check out, Santachiara Benerecetti AS, Semino O, Passarino G, Morpurgo GP, Fellous M, Modiano G (1992) "Y-chromosome DNA polymorphisms in Ashkenazi and Sepharadi Jews." In: Bonné-Tamir B, Adam A (eds) Genetic diversity among Jews: diseases and markers at the DNA level. Oxford University Press, New York, pp 45-50

In the nineties decade, most of the studies of Ashkenazic DNA ancestry markers focused on looking at the paternal lineages, the Y-chromosomes, unless the studies were specific for genetic diseases research. Population genetics that looks at ancestry focused widely on the male line of descent, and articles or research tracing the maternal lineages were difficult to find in public searches of scientific journals or even in the media.

With the new decade, more studies of the maternal lineages of Ashkenazim and non-Ashkenazic Jews are more readily available. Still, there is research yet unpublished that continuously adds new information to the research.

The question remains, who were our founding mothers and from where did they come—from ancient Israel, from Western Asia, from Europe or are we what we appear to be—a mosaic of Jewish communities that merged and intermarried over the centuries? Is there yet to be found an unbroken line yet found that links us to any one ancient civilization, or is there? Or are we mixed with the memories of expansions and migrations over thousands of years?

The mtNDA reveals the haplogroups found among non-Jews with similar origins predating any organized religion. We females are part of the K mtDNA haplogroup clan and the H mtDNA haplogroup clan that gave rise to millions of Europeans, West Asians, and Middle Eastern peoples. What new information. Are Jews a race?

The answer is they are part of the human race. The debate will go on regarding how many genes tie Jews to their ancestral palm-latitudes home in the Levant and before that in the Fertile Crescent. Is there unity in diversity? Debates on origins happen. It's more political than religious. Anyone who takes Judaism seriously can become a Jew, have a child and rear it as Jewish, and so on for thousands of years since Abraham smashed the idols in his dad's sculpture shop in ancient Sumeria. People can convert and have children who then have Jewish mothers to pass on the religion in turn to their children.

If Jews weren't linked to the Middle East, why would caricature stereotyped images portray them in cartoons with typical Semitic features and Assyroid skulls that have ancient origins in Northern Mesopotamia and the Levant? And yet if European Jews didn't also look Northern European as well, why would they be forced to wear armbands with six-pointed stars in Germany and North Eastern Europe during World War II, unless they looked so much like Germans or other North and Central Europeans that you couldn't point them out in public as appearing different in coloring or features?

Seeking a core identity and/or tailoring foods and nutraceuticals to your genes are reasons why any person of any ethnic group wants to explore ancestry by DNA testing and also nutritional genomics testing to have a genetic profile. You can put the profile in a time capsule for future generations curious about your genetic profile. You'll learn about your ancestry in ancient times, where matches are now, and with a nutritional genomics profile will find out how to get a handle on your risks and tailor what you eat to help your own health. Your signature is in your genes.

The feeling you belong to a certain group of people who wrote certain books of wisdom in ancient times is powerful. You want to belong to a family, a family within the family of humankind.

You want to celebrate holidays with others celebrating the same holidays. Jews come in all colors and in all types of DNA. And yet, there is this paternal lineage link to the Lebanese for the Ashkenazim and the Sephardim. Now what about our founding mothers? Like my former teacher once remarked, "How could a young, Jewish merchant from Baghdad or Persia traveling to Poland or to the Jewish communities in the Rhine Valley in the ninth century resist marrying a redheaded or blonde 'shiksa' goddess on the road to Speyer?" The answer could be, more genes outflowed from the Jewish males into the general local population than came in and remained.

So the origin of Ashkenazic communities will always be up for debate and theory. The details remain in the genes that reveal ancestral lines. How would historians account for the Ashkenazic Levites some, but not all, of which have genes close to Sorb males? Did Sorb males convert to Judaism to escape oppression from the Germans who invaded their Western Slavic territories to convert them to Christianity and Germanize them at a time when they were pagans resisting Christianity and the German language?

Did they team up with the Jewish communities living in East Germany in those early medieval centuries, perhaps around the year 800 CE? All scientists know is that the many Jewish communities founded independently eventually merged as more and more people flowed east into Poland upon invitation from the Polish princes. The Ashkenazic families migrated from Germany, France, and Hungary eastward. By 1000 CE, most Jewish communities were already founded and scholarship began to flourish, such as the writings of Rashi.

Who were these Ashkenazic Jews of Germany and France? They were the Roman Jews, the Jews who lived in ancient Rome who migrated northward from Southern Italy and Rome after the fall of Rome. They went were the trade flourished—trade that didn't require owning land. They minted coins and changed money, cut diamonds, or worked with gold. They sewed clothing and peddled wares.

They traveled and wrote books.

They were musicians and healers, rabbis and scholars. And they went were they were welcome. Some followed the northward movements of the Roman army. Sephardic Jews traveled the same routes, but instead of going north, went to Spain and N. Africa. Other Jews fleeing oppression branched into Iraq, Syria, Greece, and Persia. Some went out to Bukhara and Central Asia from villages in Iraq. Some went to India and some to Ethiopia.

Sometimes they found local women and married them, and other times, the women came from other Jewish communities. So you can't at least at this time in research make a sweeping statement of where all the women came from. You can only say they came mostly from other merging Jewish communities. Further scientific research again, will let the details in the genes speak their ancestral history. So keep updated and read the emerging articles in the human genetics journals.

You can access them online through most of the medical research engines on the Internet. Or you can visit a university library, especially a medical school library, and read the newest articles at the library as well as the latest books on Jewish genetic diversity or any other topic in genetics.

The particular field looks at population genetics and evolutionary genetics. Read the article that will take you back before there were borders or organized religions titled, "*Tracing European Founder Lineages in the Near Eastern mtDNA Pool.*" Martin Richards, Vincent Macaulay, et al. American Journal of Human Genetics, 67: 1251-1276, 2000. See where our mothers really came from. (The term "our mothers" refers to the human race.) You'll notice in the articles you read that the British often use the term "Near Eastern" and that the Americans use the term "Middle Eastern" for the same area.

Perhaps you want to find out the percentage of various races in your ancestry. How do know where to begin your journey into the past and future? What if you're a foundling, an orphan, or have no knowledge of your own ethnicity? Can a DNA test at least tell you how many races are in your recent or ancient past? What facts do genetic markers really tell you about ancestry?

If you want to start your ancestry search with DNA testing, first you take the DNA tests along with tests of racial percentages if you desire. Even your DNA has a cultural component to its molecular biology. Then you interpret the results making the complex easy to understand for yourself or your clients. Your DNA testing service can help you find answers. So can many Web sites as well as this book and other books recommended here.

Next in your family history search, you collect letters, diaries, oral history transcriptions, home sources, artifacts, memorabilia, Census research, wills-and-probate records, medical histories, land records, slave ownership records, if it applies to your or your client. Pay particular attention to social histories to fill some gaps left by lack of women's records.

Search through church, synagogue, mosque, pagoda, or temple records, vital records from the US government such as military records, social security information, and government pensions for retired government employees, employment and tax records, if any exist and are available. Check school records from elementary through college, if any, social histories, ethnic histories, and religious school records.

Go to the family history Web sites, the ships' passenger lists. I highly recommend a book for searching women's ancestry, titled, Discovering Your Female Ancestors, by Sharon DeBartolo Carmack, Betterway Books, Ohio 1998, ISBN # 1-55870-472-8. The book's subtitle emphasizes "Special strategies for uncovering hard-to-find information about your female lineage."

Marriage records often were in different languages representing the former country or languages of the ethnic group. You may need to translate a different alphabet to find a maiden name on a marriage certificate never registered, but obtained from clergy.

Then you review and analyze the records. Study the social history of the times and location of this individual.Add family history and migrations to social history, and you have the beginnings of an outline to write a biography of the ancestor as a family history.

Learn to interpret the results of your own DNA test and expand your historical research ability to trace your ancestry. "An interesting idea was expressed by a colleague from Canada, Dr. Charles Scriver," explains geneticist, Dr. Batsheva Bonné-Temir. "At a meeting which I organized here in Israel on Genetic Diversity Among Jews in 1990, Dr. Scriver gave a paper on 'What Are Genes Like that Doing in a Place Like This?

Human History and Molecular Prosopography.' He claimed that a biological trait has two histories, a biological component and a cultural component." Dr. Charles Scriver is founder of the DeBelle Laboratory of Biochemical Genetics in Canada. He also established screening programs in Montreal for thalassaemia and Tay Sachs Disease.

According to Bonné-Tamir, at the 1990 meeting in Israel on Genetic Diversity Among Jews, Dr. Charles Scriver stated, "When the event clusters and an important cause of it is biological, the cultural history also is likely to be important because it may explain why the persons carrying the gene are in the particular place at the time."

The term, "when the event clusters" refers to an event when genes cluster together in a DNA test because the genes are similar in origin, that is, they have a common ancestral origin in a particular area, a common ancestor.

"When I look at my own papers throughout the years," says Bonné-Tamir. "I find that I have been quite a pioneer in realizing the significance of combining the history of individuals or of populations with their biological attributes. This is now a leading undertaking in many studies which use, for example, mutations to estimate time to the most recent ancestors and alike."

What lines of inquiry are used in genetics? Dr. Charles R. Scriver wrote a chapter in Batsheva Bonné-Temir's book, titled What are genes like that doing in a place like this? Human History and Molecular Prosopography. The book title is: Genetic Diversity Among Jews: Diseases and Markers at the DNA Level. Bonné-Tamir, B. and Adam, A.

Oxford University Press. 1992. With permission, an excerpt is reprinted below from page 319: "When a disease clusters in a particular community, two lines of inquiry follow: 1. Is the clustering caused by shared environmental exposure?

Or is it explained by host susceptibility accountable to biological and/or cultural inheritance?

2. If the explanation is biological, how are the determinants inherited? These lines of inquiry imply that a disease has two different histories, one biological, the other cultural. One involves genes (heredity), pathways of development (ontogeny), and constitutional factors; the other, demography, migration and cultural practice.

Neither history is mutually exclusive. Such thinking shifts the focus of inquiry from sick populations and incidence of disease to sick individuals and the cause of their particular disease. The person with the disease becomes the object of concern which is not the same as the disease the person has." (Page. 319).

CHAPTER 9

▼

CONSUMERS CAN USE
GENETIC TESTS TO
IDENTIFY AT-RISK
RELATIVES

After hearing from Dr. Scriver by email, I then emailed Stanley M. Diamond. He contacted writer, Barbara Khait, and got permission for me to reprint in this book some of what she wrote about Diamond's project. It's the chapter, "*Genetics Study Identifies At-risk relatives*" from **Celebrating the Family** published by Ancestry.com Publishing.

Check out the Web site at:

http://shops.ancestry.com/product.asp?productid=2625&shopid=128.

Here's the reprinted article. Persons interested may go to the Web site for more information. I found out about Stanley M. Diamond from Dr. Scriver, since he mentioned Stanley M. Diamond's project in the book chapter Scriver wrote for Batsheva Bonné-Temir's book on *Genetic Diversity Among Jews: Diseases and Markers at the DNA Level*. Barbara Khait's chapter follows.

*   *   *   *

"In 1977, Stanley Diamond of Montreal learned he carried the betathalassemia genetic trait. Though common among people of Mediterranean, Middle Eastern, Southeast Asian and African descent, the trait is rare among descendants of eastern European Jews like Stan. His doctor made a full study of the family and identified Stanley's father as the source.

"Stan was spurred to action by a letter his brother received in 1991 from a previously unknown first cousin. Stan asked the cousin, "Do you carry the beta-thalassemia trait?" Though the answer was no, Stan began his journey to find out what other members of his family might be unsuspecting carriers.

"Later that year, Stan found a relative from his paternal grandmother's family, the Widelitz family. Again he asked, "Is there any incidence of anemia in your family?"His newfound cousin answered, "Oh, you mean beta-thalassemia? It's all over the family!"

"There was no question now that the trait could now be traced to Stan's grandmother, Masha Widelitz Diamond and that Masha's older brother Aaron also had to have been a carrier. Stan's next question: who passed the trait onto Masha and Aaron? Was it their mother, Sura Nowes, or their father, Jankiel Widelec?

"At the 1992 annual summer seminar on Jewish genealogy in New York City, Stan conferred with Dr. Robert Desnick, who suggested that Stan's first step should be to determine whether the trait was related to a known mutation or a gene unique to his family.He advised Stan to seek out another Montrealer, Dr. Charles Scriver of McGill University-Montreal Children's Hospital. With the help of a grant, Dr. Scriver undertook the necessary DNA screening with the goal of determining the beta-thalassemia mutation.

"During this time, Stan began to research his family's history in earnest and identified their nineteenth century home town of Ostrow Mazowiecka in Poland. With the help of birth, marriage, and death records for the Jewish population of Ostrow Mazowiecka filmed by The Church of Jesus Christ of Latter-day Saints (LOS), Stan was able to construct his family tree.

"Late in 1993, Dr. Scriver faxed the news that the mutation had been identified and that it was, in fact, a novel mutation. Independently, Dr. Ariella Oppenheim at Jerusalem's Hebrew University-Hadassah Hospital mad e a similar discovery about a woman who had recently emigrated from the former Soviet Union.

"The likelihood that we were witnessing a DNA region 'identical by descent' in the two families was impressive. We had apparently discovered a familial relationship between Stanley and the woman in Jerusalem, previously unknown to either family," says Dr. Scriver.

"It wasn't very long ago when children born with thalassemia major seldom made it past the age of ten. Recent advances have increased life span but, to stay alive, these children must undergo blood transfusions every two to four weeks. And every night, they must receive painful transfusions of a special drug for up to twelve hours.

"The repeated blood transfusions lead to a buildup of iron in the body that can damage the heart, liver, and other organs. That's why, when the disease is misdiagnosed as mild chronic anemia, the prescription of additional iron is even more harmful. Right now, no cure exists for the disease, though medical experts say experimental bone-marrow transplants and gene-therapy procedures may one day lead to one.

"Stan's primary concern is that carriers of thalassemia trait may marry, often unaware that their mild chronic anemia may be something else. To aid in his search for carriers of his family's gene mutation of the beta-thalassemia trait, he founded and coordinates an initiative known as Jewish Records Indexing-Poland, an award-winning Internet-based index of Jewish vital records in Poland, with more than one million references. This database is helping Jewish families, particularly those at increased risk for hereditary conditions and diseases, trace their medical histories, as well as geneticists."

Says Dr. Robert Burk, professor of epidemiology at the Albert Einstein College of Medicine at Yeshiva University, and principal investigator for the Cancer Longevity, Ancestry and Lifestyle (CLAL) study in the Jewish population (currently focusing on prostate cancer), "Through the establishment of a searchable database from Poland, careful analysis of the relationship between individuals will be possible at both the familial and the molecular level.

"This will afford us the opportunity to learn not only more about the Creator's great work, but will also allow (us) researchers new opportunities to dissect the cause of many diseases in large established pedigrees."

Several other medical institutions, including Yale University's Cancer Genetics Program, the Epidemiology-Genetics Program at the Johns Hopkins School of Medicine, and Mount Sinai Hospital's School of Medicine have recognized Diamond's work as an outstanding application of knowing one's family history and as a guide to others who may be trying to trace their medical histories, particularly those at increased risk for hereditary conditions and diseases.

In February 1998, in a breakthrough effort, Stanley discovered another member of his family who carried the trait. He found the descendants of Jankiel's niece and nephew—first cousins who married—David Lustig and his wife, Fanny Bengelsdorf. This was no ordinary find—he located the graves by using a map of the Ostrow Mazowiecka section of Chicago's Waldheim Cemetery and contacted the person listed as the one paying for perpetual care, David and Fanny's grandson, Alex.

"It turned out Alex, too, had been diagnosed as a beta-thalassemia carrier by his personal physician fifteen years earlier. The discovery that David and Fanny's descendants were carriers of the beta-thalassemia trait convinced Stan, Dr. Scriver, and Dr.Oppenheim that Hersz Widelec, born in 1785,must be the source of the family's novel mutation.

"'This groundbreaking work helps geneticists all over the world understand the trait and its effects on one family,' says Dr. Oppenheim.

"A most important contribution of Stanley Diamond's work is increasing the awareness among his relatives and others to the possibility that they carry a genetic trait which with proper measures, can be prevented in future generations. In addition, the work has demonstrated the power of modern genetics in identifying distant relatives, and helps to clarify how genetic diseases are being spread throughout the world."

For more information about thalassemia, contact Cooley's Anemia Foundation (129-09 26th Avenue. Flushing,New York, 11354; by phone 800-522-7222; or online at www.cooleysanemia.org). For more about Stanley Diamond's research. visit his Web site (www.diamondgen.org).

Thalassemia is not only carried by people living today in Mediterranean lands. The first Polish (not Jewish) carrier of Beta-Thal was discovered in the last few years in Bialystok, Poland. Stanley Diamond met with the Director of the Hematology Institute in Warsaw in November 2002, and the Director of the Hematology Institute in Warsaw indicated that they now have identified 52 carriers. Check out these Web sites listed below if the subject intrigues you.

"Genealogy with an extra reason"...Beta-Thalassemia Research Project.
http://www.diamondgen.org
JTA genetic disorder and Polish Jewish history
www.jta.org/page_view_story.asp?intarticleid=11608&intcategoryid=5

IAJGS Lifetime Achievement Award
http://www.jewishgen.org/ajgs/awards.html
Jewish Records Indexing—Poland
http://www.jri-poland.org

CHAPTER 10

▼

MOLECULAR REVOLUTION NUTRITIONAL GENOMICS, PHENOMICS, AND ARCHAEOGENETICS

Geneticists today are making inroads in new areas such as phenomics, nutritional genomics, and ancestral genetics. Batsheva Bonné-Tamir, PhD, http://www.tau.ac.il/medicine/USR/bonnétamirb.htm or http://www.tau.ac.il/medicine/at Tel-Aviv University, Israel, is Head of the National Laboratory for the Genetics of Israeli Populations (with Mia Horowitz) and Director of the Shalom and Varda Yoran Institute for Genome Research Tel-Aviv. She is also on the faculty of the Department of Human Genetics and Molecular Medicine, Sackler School of Medicine.

Dr. Bonné-Tamir states that "One of my most impressive conclusions from the advancement in the last few years and the accumulation of knowledge in the fields of genetics and medicine, is the molecular revolution based on immense sophistication of lab techniques. This is really responsible for the recent increased emphasis on the human-socialanthropological aspects that affect biological diversity."

Bonné-Tamir explains, "At a meeting in 1973, in my paper on Merits and Difficulties in Studies of Middle Eastern Isolates, I said that 'The Middle Eastern isolates have emphasized again the fertile and necessary interrelationship between history and genetics.'"

Do historical events influence genes? "Comparative studies in population genetics are often undertaken in order to attempt reconstruction of historical and migratory movements based on gene frequencies," says Bonné-Tamir. "The Samaritans and Karaites offer opportunities in the opposite direction, for example, to learn the influence of historical events on gene frequencies."

In another paper in 1979 on Analysis of Genetic Data on Jewish Populations, Dr. Bonné-Temir wrote that "Our purpose in studying the differences and similarities between various Jewish populations was not to determine whether a Jewish race exists. Nor was it to discover the original genes of 'ancient Hebrews,' or to retrieve genetic characteristics in the historical development of the Jews.

"Rather, it was to evaluate the extent of 'heterogeneity' in the separate opulations, to construct a profile of each population as shaped by the genetic data, and to draw inferences about the possible influences of dispersion, migration, and admixture processes on the genetic composition of these populations."

In 1999, Dr. Bonné-Temir organized an international symposium on Genomic Views of Jewish History. "And unfortunately, the many papers presented were never published," says Bonné-Temir.

Molecular Genealogy Research Projects

Certain mtDNA haplogroups and mutations or markers within the haplogroups turn up in research studies of Ashkenazim when scientists look at the maternal lineages. For example, 9.0 of Ashkenazic (Jewish) women have mtDNA haplogroups that follow the Cambride Reference Sequence (CRS) (Anderson et al. 1981).

That means their matrilineal ancestry lines follow the reference sequence that all other mtDNA haplogroup markers are compared with. The CRS shows a specific sequence of mtDNA haplogroup H found in more than 46% of all Europeans and 6% of Middle Eastern peoples. You can view a table of mtDNA sequences titled "Frequently Encountered mtDNA Hapotypes" at Table 3 in the article, *"Founding Mothers of Jewish Communities: Geographically Separated Jewish Groups Were Independently Founded by Very Few Female Ancestors,"* Mark G. Thomas et al, American Journal of Human Genetics, 70:1411-1420, 2002.

The only exception is that the Ashkenazic mtDNA haplogroup maternal ancestral lines, show up at only 2.6% with one mutation away from the CRS 343 creating U3 mtDNA haplogroup instead of the H mtDNA haplogroup. The CRS is H haplogroup. Where does U3 mtDNA show up at the higher rate of 17%? In Iraqi Jews. Ashkenazi mtDNA shows up at 9.0 percent following the CRS with H mtDNA haplogroup. Yet 27.0% of Moroccan Jews have mtDNA following the CRS. So does more than 46% of all Europeans and 25% of all Middle Eastern people have MtDNA following the CRS. That's haplogroup H of the Cambridge Reference Sequence.

The table in the article mentioned above has many sequences of mtDNA listed. Ashkenazi mtDNA was compared to MtDNA in other Jewish groups— Moroccan, Iraqi, Iranian, Georgian, Bukharan, Yemenite, Ethopian, and Indian. These were compared to non-Jewish Germans, Berbers, Syrians, Georgian non-Jews, Uzbeks, Yemenites, Ethiopian non-Jews, Hindus, and Israeli Arabs. The percentage of frequencies of HSV-1 mtDNA sequences were listed in the samples from sites 16090-16365.

Looking only at Ashkenazi mtDNA, the mutations 184 and 265T show up in 2.6% of the Ashkenazi mtDNA. Yet in Jews from Bukhara, these same mutations show up in 15.2% of mtDNA.

And 129 and 223 mutations show up in 2.6% of the Ashkenazi mtDNA. Yet in Bukharian Jewish mtDNA, this mutation shows up at 12.1%. The rest of the Ashkenazi mtDNA stands at 9.0% for matching the CRS. That's H haplogroup, the most frequent mtDNA haplogroup found in Europe. Only 1.3% of Ashkenazic mtDNA has one mutation at 274. Yet 20% of Yemenite Jewish mtDNA shows this same mutation at 274.

Be aware that not every Jewish person has been tested for mtDNA or Y-chromosomes. You first have to research how significant samples are in speaking for the majority of any population. They do have scientific credibility, but you must always look at the sample size.

Men carry their mother's mtDNA but pass on to their sons their Y-chromosome. Women pass on their mtDNA haplogroup only to their daughters.

Genomic views of any ethnic group's history are important for further study. Whether you are taking the skeptic's position or the genomic view of your cultural history, biology does have a cultural component that needs to be analyzed scientifically. Finding flaws or benefits in research studies of any kind is the way to find inroads to truths. How else can facts change and knowledge progress?

Molecular genealogy has joined efforts with molecular genetics. How can this information help you in family history research? Ugo A. Perego, MS. Senior

Project Administrator, Molecular Genealogy Research Project, Brigham Young University, http://molecular-genealogy.byu.edu, says, "I believe that DNA is the next thing in genealogy—the tool for the 21st century family historians. In the past 20 years, the genealogical world has been revolutionized by the introduction of the Internet.

"An increasing number of people are becoming interested in searching for their ancestors because through emails and websites a large world of family history information is now available to them. The greatest contribution of molecular methods to family history is the fact that in some instances family relationships and blocked genealogies can be extended even in the absence of written records.

"Adoptions, illegitimacies, names that have been changes, migrations, wars, fire, flood, etc. are all situations in which a record may become unavailable. However, no one can change our genetic composition, which we have received by those that came before us.

"Currently, DNA testing is an effective approach to help with strict paternal and maternal lines thanks to the analysis and comparison of the Y chromosome (male line) and mitochondrial DNA (female line) in individuals that have reason to believe the existence of a common paternal or maternal ancestor.

"A large database of genetic and genealogical data is currently been built by the BYU Center for Molecular Genealogy and the Sorenson Molecular Genealogy Foundation. This database will contain thousands of pedigree charts and DNA from people from all over the world. Currently it has already over 35,000 participants in it.

"The purpose of this database is to provide additional knowledge in reconstructing family lines other than the paternal and maternal, by using a large number of autosomal DNA (the DNA found in the non-sex chromosomes).

This research, known as the Molecular Genealogy Research Project is destined to take DNA for genealogists to the next level."

For additional reading, please visit the BYU's Molecular Genealogy Research Project's two Web sites. Another good source of information is at www.relativegenetics.com, a company specialized in Y chromosome analysis for family studies.

What Are Your Genes Doing In That Temporary Container?

Have you ever wondered what your genes are doing in your "temporary container" before they move on and change and where they have traveled during the past 40,000 years or more? When you have your DNA tested, work with the lab and DNA testing company, and ask them to explain to you how the 25 Y-chromosome markers you had tested as a male or mtDNA as a female help you determine your ancestry or find any matches similar to your DNA in a database. Read the frequently asked questions files of DNA testing companies online. Ask questions by email.

Sons inherit their mtDNA from their own mothers. Then mothers pass mtDNA on to their daughters. Sons also inherit the Y-chromosome from their fathers, but do not pass it to their daughters. Women don't have Y chromosomes.

The mtDNA is passed down from mother to son and mother to daughter through the cytoplasm (the cell contents surrounding the nucleus) in the egg. Only the daughters pass on their mtDNA to their daughters. And sons pass their Y-chromosomes to their sons, but those sons carry the mtDNA of their own mothers.

Even though both sons and daughters are formed from the union of a sperm and an egg, only the daughters will pass their mtDNA (which is the same as their great grandmother's on their maternal side and still further back to the first founder of their mtDNA group. Only daughters will pass the mtDNA on to the next generation.

Each lab has different methods of reporting their results, but you can use DNA test results as a tool for learning family history. Join DNA mailing lists and research or ask questions of the DNA testing firms specializing in genealogy by genetics about how many mutations occur in how many generations.

Here's how you can do your own research independently of any laboratory that tested your own DNA if you're curious about Y-chromosome DNA tests. Males take Y-chromosome DNA tests to find out paternal ancestry lineages. Males also can have their mtDNA checked, but women don't have a Y chromosome. So women only can check their maternal lineages with mitochondrial (mtDNA) tests. According to Alastair Greenshields who runs Ybase at http://www.ybase.org, "Ybase is a free and open database which allows people to enter their Y-chromosome haplotype details independently of the laboratory they were tested at. Anyone can contribute to it and anyone can explore it."

The database can accept results for any of the 36 Y-chromosome markers currently in use along with other genealogical information. "Ybase is searchable for exact haplotype matches and/or near misses,"says Greenshields. "Surnames, variant spellings and other relevant names can also be searched for, which is especially useful for the genealogist wishing to locate and contact others that share their surname and have had their DNA tested."

Most researchers that recognize their Y-chromosome cannot identify an actual individual and are happy to share their results online in an effort to find their

DNA cousins and genetic roots. "Genealogical research coupled with DNA test-ing is already proving a very powerful method of substantiating ancestry," Green-shields explains. "And Ybase is sure to grow in line with this upward trend, benefiting the genealogical community as a whole."

Are male and female genetic lineages studied for different purposes? "Ybase is solely intended for Y-DNA and not mtDNA. A database of the latter would be so general and broad, given the nature of how mtDNA is passed on, as to be of little value to a researcher,"Greenshields says.

Below are a couple of explanations on Y-DNA Alastair Greenshields wrote for two different people with entirely different backgrounds who needed the whole thing explained to them. "They explain essentially the same terms but have been 'dumbed down' to varying extents. Please feel entirely free to copy them verbatim or adjust as you deem necessary," says Greenshields.

To answer my question to Greenshields on how to interpret the results of DNA tests for Y-chromosome analysis for ancestry, he's explained it well in terms that most people can understand.

"Imagine a very long rope, some of which is lying across your desk. This is the DNA strand," Greenshields says. "It just happens to be the length of rope called 'Your Y-chromosome'.

"Now look at the bit that lies across your desk and grasp the rope with both hands, about half a meter apart. This is a 'marker'or 'locus' (Latin for 'place').We'll call it DYS19.

"In between your hands, imagine that bit of rope is divided into 14 equally-spaced segments. If you look very closely at the segments, you can see that each one has a bit of writing on it, which reads TAGA.

"This is simply the DNA code for each repeat. Therefore the marker DYS19=14 repeats. Or if I ask you, 'What allele have you got for DYS19?' You can tell me '14'. (Allele effectively means the number of repeats.) For example, DYS19 has about nine possibilities (between 11 and 19).

"If you do the same at lots of different markers or loci (plural of locus), you'll get a whole series of numbers (DYS19=14, DYS388=15, DYS461=11 etc.). This is your 'haplotype'. It doesn't matter whether it is 20 or 200 numbers in length. This series of numbers is still called your haplotype.

"Now you are going to make some rope for your new son. You are pretty good at making rope and usually you can copy your own precisely, but this time you made it slightly too short. There are now 13 repeats. (Technically this is called a 'mutation' which can occur when an enzyme mis-types the DNA code). It still works perfectly well so your son keeps it and is very happy! There, you have it—

DNA in a nutshell! (Be aware that your repeats, like the stock-market, may go up as well as down, but for entirely different reasons.)

CHAPTER 11

▼

MOLECULAR GENEALOGY AND THE Y CHROMOSOME

"DNA for the use of genealogy usually requires an analysis of your chromosome," Greenshields explains. "Only males have this particular chromosome and the DNA code held within it is passed down from father to son (virtually) unchanged. Provided there is an unbroken paternal line between two males, that is both share a g-g-g-g-g-grandfather, their Y-chromosome DNA will be the same."

When the DNA is analyzed, many small sections are looked at. Presently, the testing companies look for anywhere between 10 and 26 sections or 'markers.' "At any one of the markers, the code will repeat itself, for example, 15 times," Greenshields explains. "If the marker is called DYS19, we can give the result DYS19=15."

If you analyze several of these markers, you end up with a 'haplotype.' Thus you can compare haplotypes to see if you are related. "I say 'virtually' unchanged, as the DNA can change slightly over time due to 'mutations'—small errors formed when the DNA is copied," says Greenshields. "When comparing the haplotypes from two people, this will show up as a 'mismatch'—where, for example, DYS 19=14."

These mutations are useful by themselves and occur at a fairly steady rate over time. "It also gives us the great variability that we observe over populations," says

Greenshields. "If the mutations did not occur, every male would have identical Y-chromosomes. Also, within DNA/genealogy studies, if many mismatches occur when comparing two male haplotypes, we can say that they are not related."

So you can see, DNA can be a very useful tool when comparing 'suspected' relatives. "But there is one caveat, however," Greenshields emphasizes. "You must share a surname, or have a very good reason to believe you are related. DNA alone will not identify your relatives from any other random person."

For example, you will probably share the same haplotype to at least someone in your home-town, but having a similar or same surname will raise the probabilities significantly. "DNA is a tool that should be overlaid on the existing genealogical records," notes Greenshield. "It can be an excellent way of deciding on further avenues of research, or indeed defining that there are several distinct lines within your family name."

CHAPTER 12

▼

UNDERSTANDING HLA GENES (WHITE BLOOD CELLS) TISSUE TYPING AND ETHNICITY

Can scientists test white blood cells, do some tissue typing, and come up with your ethnicity? HLA genes are white blood cells.Anthropologists look at white blood cells called by the scientific name of HLA genes to study genetic drift. If you have a research need to learn about tissue typing, you might want to read about understanding the HLA genes. Tissue typing is usually done on white blood cells, or leukocytes. The markers are referred to as human leukocyte antigens (HLA). A good starting point is to read the definitions and excerpts about tissue typing testing on the 'About' Health and Fitness Web site at: http://thyroid.about.com/library/immune/blimm25.htm?terms=Human+Leukocyte+Antigen.

"My personal view of the HLA genes and genetic genealogy is that the two will almost certainly not be mixed in the near future on a commercial basis," says Greenshield. "My reasoning is two-fold. First, the L in HLA stands for leukocyte/leucocyte (Gk leuko=white). The test involves looking at 'white' blood cells." This would involve taking a blood sample (or possibly sampling a site of infection) and a trained phlebotomist being on hand. The Brigham Young University (BYU) study used to have participants give blood (of which HLA typing is possibly involved)."

Ugo A. Perego, MS, Senior Project Administrator, Molecular Genealogy Research Project, BYU emphasizes, "We stopped using blood last summer. All our collections are now based on a simple 45 seconds rinse using mouthwash."

"Your average genealogists are only willing to swab the inside of their mouths," says Greenshields. "Even a pin prick of blood would, possibly, be too much to ask of one's suspected relatives."DNA testing for ancestry only requires a swab of felt or cotton on the inner cheek or rinsing your mouth with mouthwash.

"Secondly, the HLA system can provide information or guides to genetic disposition to disease. Given the apprehension about giving a cheek swab to be 'junk-DNA tested' for the Y-chromosome, having HLA alleles typed, compared, and possibly posted to the net would discourage all but the hardy," says Greenshields. "I believe HLA testing for genealogical studies will remain under the domain of research into genetic drift and anthropological studies."

If you ever need a tissue donor or have to get your tissue typed for medical reasons, that's when the HLA genes play a major role in tissue typing. Interestingly, when small communities are isolated for long periods of time, and bottlenecks pare down the population to only a few founders, genetic drift may occur. That's when the anthropologists and evolutionary biologists look at the HLA genes.

With some DNA testing companies offering racial percentages tests, Y-chromosome tests, and mtDNA tests, what is being done with mtDNA testing for maternal lineages? How can we trace female relatives or ancestors who leave no written records of their name or existence?

Will studies of HLA genes be used by others in various fields as research now is used by anthropologists studying genetic drift, scientists studying tissue differentiation, or physicians looking at how white cells fight infection?

Dr. Peter Reed has a PhD in Human Genetics from the University of Oxford and was a pioneer in the use of STR genetic markers in medical research.He explains here that HLA genes primarily determine how our blood cells recognize and react to other cells present in our bodies. In particular, this makes HLA genes important in how our body responds to 'foreign bodies.'

For example, the HLA genes as white blood cells, fight infection. When bacteria or viruses enter our bodies, the HLA genes are there to do battle.When organs or tissue are transplanted,HLA genes have to be considered.

They would attack the foreign tissue placed in the body. "When people talk about blood typing or tissue matching, on the whole, they are referring to determining some aspect of the set of HLA genes," says Reed. "HLA genes are perhaps the most variable of all human genes. Across the population, some HLA genes have dozens of different forms of genes or 'alleles.'"

The result is that two randomly chosen people are unlikely to share identical HLA genes. "Even within families, there is a good chance that each family member has a different set of HLA genes," says Reed.

"That's why finding a 'suitable match' for a transplant can be difficult."

It's also the reason why we all react differently to infections. On the plus side, HLA genes (white blood cells) can deal with all the different infections we get during our lives, usually without being aware of them.

Apart from the obvious medical importance of a role in responding to infection and transplantation, there is another role. It's perhaps one of the primary reasons that HLA genes are some of the most intensely studied of all human genes.

"This relates to the role HLA genes play in determining how our blood cells respond to the other cells of the body," says Reed. "In certain circumstances, some of a person's own cells are mistaken as 'foreign bodies.' These cells are responded to as if they were an infection." The technical jargon for this is called auto-immunity.

"This can result in disease. This sort of problem is believed to be one of the underlying causes many fairly common diseases, more appropriately termed 'conditions'. Such conditions include Rheumatoid Arthritis and Juvenile Diabetes," says Reed. "A connection between particular types of HLA genes and certain conditions was first recognized more than thirty years ago." Since then many connections between HLA and human conditions have been identified.

These genes are obviously very important in human health, and are often suspected as being the major genetic causes of numerous conditions. Consequently, there are a number of clinical programs where HLA genes are screened (particularly in children) to research and even determine the risk of later disease.

"One aspect of the high variability of HLA genes, is that certain types (alleles) of certain HLA genes have been found to be geographically/ethnically distributed," says Reed. "For example, some alleles of some HLA genes may be more frequent in Japan than in England. Therefore there is some possible utility in the use of HLA genes in determining ancestry from different geographical locations. However, because the HLA genes are only a small fraction of all our genes, examining HLA genes alone is not likely to be very informative."

"Because HLA genes, like almost all our other genes, are shuffled and mixed as they are passed on from parents to children, it's difficult to determine the exact set of HLA genes of even one or two generations previous," Reed says. "So they have little utility in determining recent ancestry."

There could be some utility of HLA for genealogy. "This could be so in certain circumstances," Reed explains, "but the hurdles mentioned above will need to be overcome. I'm exploring this further."

According to Ann Turner, Genealogy-DNA List Administrator, at: http://lists.rootsweb.com/index/other/Miscellaneous/GENEALOGYDNA.html, an excellent Web site for explaining HLA is located at the Web site: http://www.med.umich.edu/trans/public/hla/hla_&_you.html.

"This HLA Web site diagrams the inheritance patterns. It says HLA is on gene 6, but it means chromosome 6," reports Turner. "You also can learn about linkage disequilibrium at this site. Some genes in the HLA system are close to one another. That makes the alleles, which are a form of a gene, also linked together closely and inherited as one unit, or haplotype."

That's the original context for the word 'haplotype.' Also look at: http://www.hokkaido.bc.jrc.or.jp/laboratory/laboratory500_eng.htm, Turner notes.

Understanding Your Maternal Lineages—mtDNA

MtDNA shows ancestry passed from mother to daughter from a single common ancestor or founder. Every human owes his or her ancestry to the ultimate "Mitochondrial Eve" the first woman to walk out of Africa and head towards Yemen around 154,000 years ago, give or take a few thousand years.

People who have their DNA tested for ancestry research want their results kept private and not used against them, by employers or insurers looking for genetic risks, or by groups that use the results of DNA tests to begrudge them of anything from their rights to religious and personal choices to their core identity. DNA analysis for ancestry does not report on disease risk.

It's okay to have databases with your DNA matches so you can contact what might turn out to be someone who shared a common ancestor with you 250 years ago or thousands of years ago or more recently. People who were reared in adoptive families or foster homes would cherish the idea that they had ancestors who lived in a particular geographic area that they could visit or look to for a core identity.

Sometimes the country you live in provides the core identity and other times some people want something more personal or religious as a core identity within a core identity. For example, someone may be Jewish, but knowing they are

Sephardic, Ashkenazic, Mizrahi or another group provides an open door to sampling travel, identity, food sampling, religious and ethnic customs, history and more, perhaps a search for ancestors or relatives.

If we study mitochondrial "Eve" then we have to study Y-chromosome "Adam." We have to ask whether mtDNA diversity is higher than Y-chromosome diversity because mtDNA developed and mutated at a different rate than Y chromosomes if we look to prehistoric ancestry lines.

Usually, studies of mtDNA show either some women had a more diverse genetic history or some communities were founded by very few female founders. For example, H haplogroup of mtDNA found in a large percentage of Europeans may have begun in the Dordogne valley of what is now France and/or in northern Spain about 21,000 years ago, but before that, H haplogroup may have had an ancestor somewhere else.

That common ancestor was one woman who had at least two daughters who survived to have more daughters and who lived somewhere in the Middle East. At some point back in time, H haplogroup arose from a still more ancient common ancestor, another woman, who lived outside of Europe.

What we see now are the mutations that occurred over thousands of years since haplogroup H mtDNA reached Europe and expanded to cover today all of Europe from Iceland to the Urals. H haplogroup mtDNA today is found in places as far apart as Bashkortostan in the Urals and Iceland, Scotland, Spain, Norway, Austria, Turkey, Crete, Ukraine, Italy, and Bulgaria.

MtDNA haplogroups are classified as A, B, C, D, E, F, G H through J, K, M, N, O, P, R, and T through Z. Then some are given little sub classifications such as U1, U2, U3, U4, U5, U6, U7, and various types of U found in mostly in India. New mtDNA haplogroups are still be uncovered.M is a super haplogroup divided into various groups of M such as M1 and M11. As ancient burials are uncovered, different mtDNA haplogroups turn up that are not here today because they are very ancient and did not survive because some women had only sons and some daughters didn't survive to reproduce.

CHAPTER 13

▼

USING YOUR OWN DNA TESTS RESULTS AS GENEALOGY TOOLS

It's good to have a mentor to answer questions about your test results until you are able to do your own research on the Web. If you're a lay person, where can you learn enough molecular genetics to get a handle on DNA test results and untangle ancestral roots? If you've ever wondered why your genes are not where you thought they were supposed to be (in geographic location on a map), that topic of research is called molecular prosopography. See the Web site: <u>www. linacre.ox.ac.uk/research/prosop/prosopo.stm</u>.

Prosopography is an independent science of social history embracing genealogy, onomastics and demography. Prosopography is all about human history and genes that travel because your genes have both a cultural and a biological component. The cultural component includes onomastics which is the study of the origin of a name and its geographical and historical utilization.

Onomastics includes the study of how and when place-names were originated and used. Then there's toponymics. Toponymics is the study of names related to a place or region. See http://libraryweb.utep.edu/onomastics.html or http://www.kami.demon.co.uk/gesithas/biblio/bib08.html. And you probably know demography, is the interdisciplinary study of human populations. Demography deals with social characteristics of the population and their development. So you'd find more information on demography by researching population studies. Phenomics is the science of customizing, tailoring, and individualizing medicines and other health treatments to the total human genome of one person.

The age of one medicine or hormone fits all is gone. As a tool, phenomics also can be applied to herbal remedies, food supplements, vitamins and minerals, hormones, and other formulas adjusted to an individual's total genome. If you have a genetic risk for a certain disease, perhaps you can find out what way there is to prevent it by using phenomics as a tool for customizing your treatment or working on prevention strategies of lifestyle, diet, or medicine.

Family history DNA testing is a new way to approach biological research. Genealogy and genetics are forms of hunting and gathering that persist. First you start with transcribed oral history. We are foragers in molecular family history.

Molecular genealogy uses DNA testing (human genetics) as a tool for untangling ancestral and recent family roots. Here's an introduction to family DNA testing to be used with oral history gathering and genealogy.

Start your family history time capsule, gift basket, scrapbook, genetic genealogy, or begin a small business publicizing DNA testing for genealogy. The place for genetic genealogy is in an archives, library, museum, or good storage place.

Future generations need a DNA history of as many ancestors as they can find willing to participate and to create oral histories. Genetics is the most mathematical/statistical of the biological sciences. We have fields such as bioinformatics that combine computers and biological information. Family historians need a bridge to fill the gap between such a mathematical science as genetics and genealogy, often based on records and oral histories.

The oral history would be transcribed on acid-free paper in hard, bound copy. Photos and other memorabilia could be added. Then the basic archive would be copied onto disks such as a CD, DVD, or other, stored in a computer and on video and audio tape.

Another copy would be saved as a multimedia presentation with text, sound, voice, photos, illustrations, and video/audio and saved on a disk to be played on screen with a home entertainment player or in a computer.

You could put a smaller file online on a Web site. This molecular biography would represent not only the life of a person, but a history of the person's DNA test results, racial percentages, ethnicity, if known, and anything else about the DNA sequence as far as geographic location or even medical history, if desired, in a more private file for relatives. This is where genetics joins with genealogy.

We not only have a family history to archive, but now a genome, or at least a record of the matrilineal and patrilineal ancestry by DNA. We have the markers and the sequences. The idea is to learn enough about DNA testing and genealogy to understand what those sequences and markers mean.

What can we learn about ancestry through the mitochondrial DNA (for women and men) the Y chromosome only for men, and other markers on the genome? What should we look at to view the percentages of races such as Native American, African, East Asian, or Indo-European (Europe,Middle East, and India)?

What do these sequences tell us about our ancestry? If there's no such thing as race, what geographic locations of our ancestors are we viewing back in time when we look at the genetic markers?

What dates are we looking at—a few generations ago or 21,000 years? What do our transitions and mutations mean over a long span of time? What foods, medicines, therapies, and climates are best for our customized, individual molecular profiles?

How do we read and interpret those genetic markers? Where does genealogy and oral history fit in? Family history—genealogy—now has joined up with molecular genetics and evolutionary anthropology. And included with genealogy is the tradition of transcribing and recording oral history, diary journaling and restoration, time capsules, biography, scrap booking, videography, and photography.

The genome has reached the genealogist. Family history today is multimedia and molecular, historical and futuristic. "Progress in our knowledge of the genome and of its function has been extremely rapid since the development, in the mid-eighties, of the Polymerase Chain Reaction," says Professor of Genetics, Guido Barbujani, (Department of Biology, University of Ferrara, Italy.) Dipartimento di Biologia, Universita' di Ferrara via L. Borsari 46, I-44100 Ferrara, Italia. See his Web site at: http://www.unife.it/genetica/Guido/Guido.html.

Dr. Barbujani's fields of interest include human population and molecular genetics and evolution, and I've read many of his articles in the various journals of genetics and research books, such as Archaeogenetics: DNA and the population prehistory of Europe published by the McDonald Institute Monographs.

"By that method, minimal quantities of DNA can be studied, which has opened the field for a number of previously hard-to-imagine applications, ranging from gene therapy to the prediction of interactions among genes, from the sequencing of entire genomes to the retrieval of DNA sequences from extinct organisms," Barbujani explains.

"DNA technologies proved so powerful that people tend to forget about their limitations. Still, limitations exist, especially in the field of genealogical reconstructions, and future technical advancements are unlikely to be of great help.

"Consider this: Each of us has two parents, four grandparents, eight grand-grandparents, and so on. In principle, only ten generations ago (around 1750 AD) we had 1024 different ancestors. In fact, chances are our ancestors were less than 1024, because consanguineous marriages likely occurred at various stages. But even if we had only 200 independent ancestors ten generations ago, each of them contributed to our 30,000 or so genes.

"On the other hand, only one of them transmitted to us her mitochondrial DNA and, if we are males, from only one of them did we inherit our Y chromosome," Barbujani reveals. "The other 198 or 199 ancestors' contributions to our genotype are of course equally important, but there is no easy way to figure them out."

"Indeed, at every generation recombination created new associations of genes along our chromosomes, except for the mitochondrial DNA and for part of the Y chromosome, which do not recombine. In this way, traits of DNA coming from different ancestors have been assembled in a mosaic that cannot be disentangled a posteriori, in which each piece has a different, and possibly very different, origin. In short, it is an illusion to think that our mitochondrial DNA (or our Y chromosome) may allow us to understand our family history.

"These are small parts of our genome, and hence contain information on but a small bit of our biological history," says Barbujani. "Other ancestors have transmitted to us many more genes than the ancestors from whom we inherited our mitochondrial DNA, and they may have come from different parts of the world."

"That may sound frustrating to some, but population genetics has something important to tell us in this regard. Population histories are much easier to reconstruct than individual histories, because chance phenomena have a much greater impact on the latter.

"When a large number of individuals are jointly analyzed, rather robust evolutionary inferences may be drawn, even if some members of the sample have had an unusual family history. By combining measures of genetic diversity, among populations and among individuals, with the evidence coming from mitochondrial and Y-chromosome genealogies, population geneticists have shown very clearly that each population contains a large proportion of all humankind's alleles, around 85 percent, on average.

"This finding has several implications. One is: should most humans disappear because of some global catastrophe, and should only one community survive, the loss of genetic diversity would be very limited, around 15 percent. That might or might not be reassuring, but is true.

"Secondly, although many tend to think that humans come in clear racial clusters, that is not true; if, on average, populations contain 85 percent of the global human diversity, two individuals from very distant localities can be just 15 percent more different genetically than members of the same population (unless the latter are relatives, of course).

Third, if genetic diversity is so high among members of the same population, the only possible explanation is that those populations incorporated, through time, contributions from other populations at a rather high rate.

"In other words, our ancestors spent most of their evolutionary time in communities connected by extensive migratory exchanges, and not in isolated groups. Through migration, alleles of African, Asian and European origin ended up all over the world, and no biologically recognizable race evolved in our species. Therefore, it is impossible to define our origin by studying our DNA, but if it were possible, we would probably find that our roots are spread over much of the world.

"As Jonathan Marks remarked, today convincing people that there is no such thing as a human race is probably as difficult as, in the 17th century, to convince people that the earth rotates around the sun and not vice versa. However, this is a scientific fact, and perhaps the single most significant result of human evolutionary studies. Everybody can tell a Nigerian from a Japanese person, but if we move from Nigeria to Japan we shall never find a sharp boundary separating two well-distinct groups.

"Rather, we shall notice that the genetic features of people change continuously, in a gradient, and that each community harbors substantial biological differences among its members. The best way to summarize these concepts, I think, is by a slogan invented by the French anthropologist André Langaney: Tous parents, tous différénts. We are all relatives, and we are all different."

The Cohen Modal Haplotype and the Armenian Modal Haplotype

If you want to look for signatures of Middle Eastern origin in Ashkenazic males you first look for repeat patterns that support the hypothesis that haplotypes with high DYS388 repeat. If they repeat, then you can hypothesize that there is an origin in the Middle East. The DYS388 repeat becomes one of the signatures of ancestry of Middle Eastern or Southeast Asian origin (Bradman et al. 2000).

For example, in Armenians, the DYS388 marker of the Y-chromosome repeats 15 times. In Georgians the repeat is 12 times. Does it repeat in the majority of Ashkenazic males? The Armenian modal hg2 haplotype is shared with other Middle Eastern countries, but the Armenian modal hg1 haplotype is a one-step relative of the English modal haplotype 3 and the Frisan modal haplotype 50 of the hg1 haplotypes. The Armenian hg1 haplotype also is the Turkish modal hg1 haplotype.

Scientists know that the Cohen Modal Haplotype shows up in non-Jews in Hungary, Italy, and Armenia, but there's also the modal hg2 haplotype in Turkey and Azerbaijan. Armenians share haplotypes that occur frequently in Jews and in other populations. The Cohen Modal Haplotype is strikingly similar in Sephardic and Ashkenazic Y-chromosomes—the paternal lines. There are regional DNA differences in Armenia. And Armenia had a vibrant Jewish community in medieval times that disappeared several hundred years ago.

The community existed contemporaneously with Jewish communities at the tail end of the era of the Khazars (a few were still around in 1200 after Khazaria was destroyed in 965 CE). Jewish communities flourished in the 13th century in neighboring Georgia and also in the Caucasus, in Iran, Azerbaijan, Daghestan (original homeland of the Khazars), the Crimea, (the refugium for Byzantine Jews and people fleeing Germany and Hungary). The Armenian Jewish community was a neighbor to the Jewish communities of the Ukraine, where Jews from Germany and Central Europe joined existing communities of Jews from Anatolia and Greece fleeing to the Ukraine from both West and South. Byzantine Jews lived in the Ukraine and the Crimea, as well as Khazars who may have joined the nearby Jewish communities. What is known is that Jewish men traveled far and wide within existing Jewish communities. Medieval Armenian Jews had Hebrew names. The city of Eghegis contains 62 medieval Armenian Jewish gravestones dated around the year 1266 CE with Aramaic and Hebrew inscriptions.

The medieval Armenian Jewish community was small—perhaps consisting of only about 150 people, according to Frank Brown, writing in the Jerusalem Report ("Stones from the River"). The stone inscriptions contain dates ranging from the middle of the 13th century to 1337.

See the Web site on Jewish Armenian communities in medieval times at: http://www.sefarad.org/publication/lm/045/4.html.

DNA of the Levites

The Levites are different. Levites, unlike Cohanims, have some Y-chromosomes in three different groups showing a heterogeneous, a diverse, variable origin. Some contemporary Levites may not be direct patrilineal descendants of a paternally related group. There is a term, "the Ashkenazic Levite Modal Haplotype." See the Scientific Correspondence section of Nature, Vol. 394, 9 July 1998.

Studies look at the marker DYS 388 in the Y-chromosome to see its repeat patterns. Scientists also use a model as a method to estimate the coalescence time of Cohanim chromosomes and in certain studies have dropped DYS388 in analyzing Cohanim chromosomes to estimate coaslescent time. Ashkenazic and Sephardic Cohanim chromosomes have been dated to an estimate of 2,684 to 3,221 years before present.

The earlier date goes back to the Exodus and the latter date goes back to the destruction of the first Temple in 586 BCE. Studies conclude that the origin of "priestly Cohanim Y-chromosomes" originated sometime during or right before the Temple period in the history of Judaism. There's always uncertainty because the mutation rate varies.

In some of the older anthropology textbooks, Jews and Armenians have been classed together as "Armenoid or Assyrian" in characteristics such as skull shape and features.

What actually happens is when a broad-skulled person from any country marries a narrow-skulled person, the child sometimes is born with an "Armenoid" skull, that is a flattened occiput, narrow skull, long or oval face, and a convex nose with complexions varying from fair or freckled and fair to sallow and/or swarthy, and hair color from blonde or red to brown or black.

People with these features can come from *any* country in Europe or the Middle East, from along the Adriatic to Armenia and northern Mesopotamia, Syria, Western Greece, Italy, or anywhere else in Western Eurasia. These characteristics for decades had been assigned to the "Armenoid" peoples.

Tall versions of the type were in the past assigned to people living along the Adriatic. Armenoids in the past also had been labeled "Assyroids." Ashkenazim and Armenians had been grouped together before DNA testing was used by physical anthropologists, population geneticists, or archaeogeneticists.

For more information on DNA studies of the Levites see the article, "The Origins Of Ashkenazic Levites: Many Ashkenazic Levites Probably Have A Paternal Descent From East Europeans Or West Asians," Bradman, N1, Rosengarten, D and Skorecki, K21, The Centre for Genetic Anthropology, Departments of Biology and Anthropology, University College London, London, UK. And 2 Bruce Rappaport Faculty of Medicine and Research Institute, Technion, Haifa 31096, Israel and Rambam Medical Center, Haifa 31096, Israel. (See the Bradman Index at http://dna6.com/abstracts/bradman.htm which is online as part of an index to a list of abstracts and posters at http://dna6.com/abstracts/index.htm

Here are some definitions you might want to peruse before we go into the next chapter on personalizing family history records with the results of DNA tests. If you are a historian or genealogist, it would be useful to be able to discuss possible DNA testing with your clients. Molecular tools to family history research open doors to new subjects.

Useful DNA Definitions for Historians and Genealogists Interested in Molecular Anthropology/Archaeology

*Genome. A person's genome is one set of his (or her) | genes. The human genes, which control a cell's structure, | operation, and division, are located in the cell's nucleus. The | full human genome (estimated at 50,000 to 100,000 genes) is present in every cell-nucleus, even though many genes are| inactive in cells which have some specialized functions (the| "differentiated" cells).

*Genes and Chromosomes. Genes are composed of segments of DNA. In normal cell-nuclei, the DNA is distributed among 46 chromosomes (23 inherited at conception from a person's father, and 23 from the mother). Each chromosome consists of one very long strand of DNA and numerous proteins, which are required for successful management of the long DNA molecule. The longest chromosomes each "carry" thousands of genes. Every time a cell divides, the cell must duplicate the 46 chromosomes and must distribute one copy of each to the two resulting cells.

*The Code. The DNA of each chromosome is composed of units—nucleotides" of four different types (A, T, G, C). These nucleotides are linked to each other in linear fashion. The sequence of the four types of nucleotides is critical, because the sequence produces the "code" which (a) determines the function of each particular gene, (b) identifies the gene's start-point and stop-point along the DNA strand, and (c) permits certain regulatory functions. The code of the human genome consists of more than a billion nucleotides.

*The Mitochondrial DNA (mtDNA). Outside the nucleus, human cells also have some "foreign" DNA located in structures called the mitochondria. This small and separate set of DNA does not participate in the 46 human chromosomes, and is not part of "the genomic DNA." The mitochondria are inherited from the mother.

These genetic term definitions are from the book titled: Confirmation that Ionizing Radiation Can Induce Genomic Instability: What is Genomic Instability, and Why Is It So Important? John W. Gofman, M.D., Ph.D., and Egan O'Connor, Executive Director, CNR. Spring, 1998. The excerpt of definitions from Dr. Gofman's essays, such as what is quoted above is reprinted with permission. See excerpts from the book at the Web site at: http://www.ratical.org/radiation/ CNR/GenomicInst.html.

All the definitions from Dr. Gofman's essays are available for eproduction in other publications. Please do cite the title and above URL so people who wish to study the complete work can do so. For more information see the Web site at: CNR page (http://www.ratical.org/radiation/CNR/). For further information, contact the publisher, David Ratcliffe, "rat haus reality press" at: http://www. ratical.org/rhrPress.html.

CHAPTER 14

▼

PERSONALIZING RECORDS
AS TIME CAPSULES

Demography will take you into old city directories. When you have finished your genetic profile tests for nutritional genomics and selected the prescribed foods and nutraceuticals based on your gene expression, you also can choose a DNA for ancestry testing company that provides a database for you to find DNA matches to your mtDNA or Y-chromosome or let you take racial percentages tests.

Then you can find your match and interview the individual or chat and create your "DNA Match" time capsules of your own, a journal, audio tape, video, or other archive where your family and the other family shares ancestral history events in an oral or transcribed archive or record. Many DNA matches of yours will want to remain private. Some may ask to be contacted or contact you for correspondence or to match possible distant relatives.

What usually happens is that people leave messages on databases and may or may not contact one another. If you're looking for distant cousins, you have the chance here to find ancestors by DNA matches. However, with only Y-chromosome or mtDNA tested, you'd need a certain number of matches on the Y-chromosome to possibly be related. Some companies will test for 25 markers. Ask your DNA testing company about the chances you are related to someone with matches on 25 of your Y-chromosome markers. Use records and names also from the past. This is where historical records, city directories, or the genealogist comes into the picture.

Or you can work with family diaries of events uniting ancestors. Diaries and DNA testing personalize family history records. DNA family databases and scrap booking are more valuable when linked together. It's time to compile a written and illustrated family tree time capsule in any or all of various media—print, pictures, video, audio, for the Web, in a scrapbook form of stories, anecdotes, experiences, photos, journal writings, and memorabilia that also includes DNA testing.

DNA test results and autobiographies also may highlight the important events that you want remembered. In the future, families may be able to archive the sequences of their entire genome—all their genes—into a database to be kept along with photos, video, audio, crafts, and memorabilia marking the life experiences, events, rites of passage, and highlights focused around a central issue.

You can piece together records of women's clubs, diaries, and DNA tests and look at your maternal lineages. String together military pension or service records, village societies, or Census records and city directories, and link your paternal lineage from voter lists and court records to Y-chromosome test results.

The file or database could be passed onto other family members and the genome given to health care professionals to customize therapy or treatment, tailor foods and vitamins or supplements, or create a living video biography. For now, whole genome DNA testing is expensive, and what are affordable include the matrilineal and patrilineal lineages and the percentages of races.

For family historians, there are also the surname databases and message boards on the Web. And for tracing female ancestors, there are marriage certificates, birth records, court records, church and synagogue records, records of teachers, factory workers, census records, social and immigrant records, women's publications, clubs, insurance records, deeds and wills, and other records that reveal more than the words, "and wife."

The 1850 US Census was the first census to name all members of a household, their birthplaces, and whether married within the year. By 1870, the US Census asked whether one's parents were foreign born. And a decade later, the 1880 US Census named relationships to the head of household.

By 1900, the US Census included the number of years married, number of children born and who of the children were living. They also added immigration and naturalization data. Native Americans had special censuses. So you can also look at school censuses and state and local censuses as well as city directories published until 1976.

Years ago when people had no phones, the city directory was one way to locate families. In creating a "family history memorabilia time capsule" or database that includes DNA testing, you can include small crafts such as braided hair embroidery as art, craft, needlework, preserved clothing or wedding gowns restored and wrapped.

You'd index where the craft work or clothing is located, and the index or list of memorabilia would go into your scrapbook, database, or time capsule. What new item that you might add, would be the family's DNA, the male and female mtDNA and the male Y chromosome, plus a racial percentages DNA test.

Records may be copies and stored in many ways—as Web sites, printed books, diaries, on CDs and DVDs, as video and audio tapes, as oral history transcripts printed out in text, or as a photo scrapbook with captions. Or you can create a multimedia presentation combining text, voice, video, photographs,music, and commentary.

You would ask as many relatives as you can find to lightly rub a felt tip or pad, cytobrush, or a swish of mouthwash in their mouths and send the samples packaged and labeled separately to a DNA for genealogy testing company. With all those entries in your time capsule, your descendants (or your clients' if you do family history research) will have a better idea of who any particular family was as people (rather than some anonymous photos).

You can even search antique stores and flea markets and the people listed on the Web. Some genealogists rescue old photos and are listed on the Web. Check to see whether any photos found in certain locations might be your family members or those of your clients if you are a genealogist or family history researcher.

So you wipe a felt swab or small brush across the inside of your cheek and mail it to a DNA testing company. Or you swish a type of mouth wash and expectorate it into a container and mail it back to a DNA testing company emphasizing testing and/or researching genealogy by DNA analysis.

Some of these companies may also have a division that tests DNA for forensic purposes. Some companies have their own laboratories for DNA testing. Other firms may give the DNA to a laboratory or contract with a university research or teaching laboratory for actual testing. Some firms would send you the results with information on how to further research your ancient lineages or genealogy for more recent DNA matches. When testing for nutritional genomics, you might be dealing with a hospital laboratory, a private company, a nutritional genomic testing company, a university research laboratory, or even a DNA bank started by a support group of families of patients with an interest in DNA testing for a specific inheritable condition. Sometimes genealogy societies might show interest in starting their own DNA bank for a particular ethnic group, extended family, or genetic condition. Some private DNA banks are started by parents of children with a specific genetic condition for the purpose of collecting DNA for research done by medical schools and other research teams on that particular genetic condition.

For DNA ancestry or racial percentages tests, what comes back to you in the mail several weeks later are a print out, perhaps a CD, or a mailing on paper and/ or email table of some of your DNA sequences. Now it's your job to find out what the sequences mean.

Most companies have frequently answered questions message boards and some firms will email you answers to your questions. Other companies may offer to store your DNA or a certain length of time. So check with the company on what it offers regarding DNA testing and genealogy questions answered.

"The state of identity testing is such that people should have a specific hypothesis that they want to test," says Harry Ostrer, M.D,Professor of Pediatrics, Pathology, and Medicine Director, Human Genetics Program New York University School of Medicine.

"Do Susan Smith and I share a common matrilineal ancestor?

"Do Jeffrey Jones and I (if male) share a common patrilineal ancestor?" Dr. Ostrer asks. "Hoping to discover something unanticipated is unrealistic. It is very unlikely that amateur genealogists will discover that they had Amerindian ancestry unless they had a strong reason to expect so.

"The problem of course is that many people are searching for roots and hoping that genetic testing will fill the gap for absence of familial oral histories," Dr. Ostrer explains. "Unfortunately, there are no shortcuts to the work of the genealogist. With the Internet, email and a heightened awareness of genealogy, the tools—word-of-mouth and access to vital records—are more accessible."

According to Dr. Harry Ostrer's article, "A Genetic Profile of Contemporary Jewish Populations," in Nature Reviews/Genetics, Vol. 2, November 2001 (Science and Society), p. 895, Macmillan Magazines Ltd, "The Ashkenazi Jewish population in Eastern Europe expanded rapidly, growing from an estimated 10,000-15,000 people in 1500 to 2 million in 1800 and 8 million in 1939 (REF.34)."

Compare that profile to the group of people you're researching. What's the individual's genetic profile? How does it compare to the oral history profile or the written record profile? Compare social or medical, ethnic, or industrial historical records from institutions, hospitals, ethnic associations, industrial employment records, military service records, old photographs, and other historical records.

Consider the group of people you may belong to on one or both sides of your family. Then find out about the genetic history of your people in the same way as you research the genealogy or family history—through reading articles on the molecular genetics history of your ethnic group. You can also talk to relatives and even trace each ancestor's medical history as far back as oral history or written records take you.

How can you use DNA testing information together with oral histories, diaries, military and court records? You can do online genealogy searches. Explore by reading and interviewing scientists or listening to tapes or videos.

Research your own genetic origins or any ethnic group you want to study. If you're not Jewish, the echniques are universal. Learn about how DNA techniques and molecular genealogy meet. Apply the methods to your own family history research.

Dr. Ostrer of New York University School of Medicine, Human Genetics Program conducted genetic analysis of Jewish origins. See the Web site at: http://www.med.nyu.edu/genetics/jewishorigins.html

According to the Web site, "The next step in Jewish genetic demography will be to understand the patterns of Jewish migration that formed the historical communities. Clearly most of these communities no longer exist, but their genetic structure can be discerned by studying the DNA of their descendants."

Consider looking at any other family memorabilia such as photographs or paintings, even crafts made in the past. Oral history, DNA testing, and genetic history work together with written, medical, and oral family history. Consider the times and background of the dates.

Search the industrial revolution era or before up to the present. What conditions did the person you're searching live under—agricultural or industrial? Was the family confined to a tiny apartment in an urban setting or on a farm?

Once you receive the results of DNA testing, you'll have a collection of equences. Now is your chance to learn how to interpret those equences by asking questions on the DNA mailing lists and reading up on the subject of what the sequences mean in plain words.

To find out where to get your questions answered about interpreting the sequences, first check with the company that tested your DNA as various labs use different markings. Then contact the message board at Roots Web.com and subscribe to the genealogy digest known as GENEALOGY-DNA-D.

Subscribe to the mailing list and receive frequent email, or just the digest, or read the messages at the Web site. There, you may ask your questions about how to interpret your own sequences or others. To subscribe to GENEALOGY-

DNA-D, send a message to GENEALOGY-DNA-D-request@rootsweb.com that contains in the body of the message the command subscribe and no other text.

No subject line is necessary, but if your software requires one, just use subscribe in the subject, too. To contact the GENEALOGY-DNA-D list administrator, send mail to:

GENEALOGY-DNA-admin@rootsweb.com. Ask the company who tested your DNA to tell you how to interpret your sequences. Most companies have a frequently asked questions section on the Web site, and others ask to be emailed questions.

Are you interested in researching, collecting information, scrap booking, genealogy, or writing about family histories or your genetic history? How can you create a time capsule for future generations of printouts of part of your DNA or mtDNA or Y chromosome sequences?

What will future generations do with this information? Can it help unite people who are distant or close relatives or those with the same common ancestor in the very distant past? Family historians and genealogists now have a new branch of genealogy to learn—molecular family history.

If you're interested in DNA and Jewish genealogy, write to:

JewishGen, Inc. 2951 Marina Bay Dr., Suite 130-472 League City, Texas 77573. On the Internet, the Web site is at: http://www.jewishgen.org/. There is a special interest group on Jewish DNA research and genealogy. It's called Genealogy by Genetics, and the Web site is at: http://www.jewishgen.org/dna/. You can subscribe to the mailing list.

The Genealogy by Genetics special interest group of the Jewish Genealogy Web mailing lists on the Web have a partner, Family Tree DNA (FTDNA) that tests your DNA and if you select,racial percentages. The DNA testing partner maintains databases. They will be integrating the FTDNA database library with existing JewishGen databases to provide users with the ability to connect with lost branches of their families.

There are several excellent books written on how to interpret DNA tests for people without a science background. See the bibliography at the back of this book. I particularly found Alan Savin's book very informative in bringing together DNA testing knowledge to genealogists. Alan Savin of Maidenhead, England, is author of the 32-page book, DNA for Family Historians (ISBN 0-9539171-0-X). See the Web site: http://www.savin.org/dna/dna-book.html.

This excellent book that I highly recommend explains and explores in layman's language how family historians-genealogists can use DNA research and test results for family history research. The book also has case studies and makes genetic theory easier to understand by those without a background in genetics. It discusses the practicality of DNA testing for family historians as genetics joins genealogy. And it includes discussion of some of the problems of using DNA testing as a tool for family history research.

What I like about this book is that it's written at a reading level that is clear to understand without a science background. And the reader will find a good introduction, historical background, explanation of DNA fingerprinting, mitochondrial DNA testing, Y chromosome DNA testing (for males), collecting and analyzing DNA, future developments, and an excellent bibliography that includes Web sites, magazine articles, and books. So when I contacted Alan Savin by email, he related to me his story of how he introduced genetics into genealogy.

"I believe I was the first family historian in the world to use DNA for enealogical research back in 1997," says Alan Savin. "I originated the phrase 'genetic genealogy'. Realizing its potential, I wished to share this with everybody, hence the writing of the book. It is still selling well, especially in the USA, with orders being received worldwide.

I have been approached recently for the book to be translated into German. It has been well received and recommended by a spectrum of reviewers from many genealogical publications, DNA testing companies themselves, e.g. Family Tree DNA and the media, e.g. the BBC."

"As stated in its introduction 'one of my primary aims is to explain this area of genetics in a language easily understood by a genealogist or any lay person'. Further books are planned in the series to develop the themes."

Savin says, "I could be said to be the father of genetic genealogy and I have seen my idea grow with the help of others. I keep a close watch on its development, behind the scenes, and look forward to seeing the science reach its maturity."

Aside from reading books on DNA, what else can a family historian do when there are no records to be found? Hobbyists and professional genealogists who wish to extend their family trees by confirming a link where no conventional source records exist would be interesting in having their DNA tested.

Sometimes DNA tests may be used to determine whether a person is part of a larger group of people: for example, Jews will be able to confirm they are of Cohanim lineage. DNA tests are excellent for individuals who want to perform surname-based family tree reconstruction projects.

An excellent article containing another version of the quote below (used with permission) titled, "Tangled Roots? Genetics Meets Genealogy" by Kathryn Brown appeared in the publication, Science, 1 Mar 2002.

Commenting on the role of DNA testing companies, Peter Underhill, a molecular anthropologist at Stanford University admits,"My concern is that people comprehend the relatively low level of resolution offered by these tests. Because the tests analyze relatively few markers along Y DNA or mtDNA, millions of people may share a given molecular profile. I think these companies have a role to play, as long as the science isdone well."

Terry Melton, PhD is President/CEO/Laboratory Director of Mitotyping Technologies, LLC, 1981 Pine Hall Drive, State College, PA 16801. "The most important contribution of this science to genealogy is the ability of mtDNA to trace the maternal line long distances throughout a family tree," says Dr.Melton."My favorite pedigree is one from a paper by Sigurdardottir (American Journal of Human Genetics 66:1599) showing fifteen generations of an Icelandic family where living individuals typed from extreme tips of the family (whose ancestral female dates back to 1560) have the same mtDNA profile.

"In addition, mtDNA can sometimes be used to illuminate ethnic ancestry (in a very general way). Mitochondrial DNA types are correlated with the region of the world where the ancestral lineages originate.

There have been dozens of scientific papers written on this subject.

"Unlike Y chromosome typing, which should follow the patriline (and family name), mitochondrial DNA is more difficult to correlate with recorded genealogy, since female names are lost in marriages," Melton explains. "However, the possibility remains that if a family can locate two (even very distant) maternal relatives in their tree, the mtDNA typing can confirm the matrilineal connection."

In addition to checking DNA test results with databases and tables on the World Wide Web or in other records, the family historian can compare results and read further about DNA to learn more.

Family members who have their DNA tested also can also track lineages and more recent genealogy by looking at the tell-tale clues that old,antique photos offer as well as use old and new city directories that list people who may not even have had a telephone.

Marjorie Rice rescues old family photos from antique stores and flea markets using the skills and sharp eye of a genealogy researcher to get them back into the hands of family members.An article about her work is on the Web at: http://www.ancestry.com/library/view/news/articles/6590.asp.

She looks on the back of the photos to see whether there are family names and/or photographer imprints on the front. She posts the names and locations on surname message boards on the Internet. To date, she has restored 409 photos to family members. See article about her work at: http://www.ancestry.com/library/view/news/articles/6590.asp.

Besides putting dozens of family photos from the early 20th century in my own computer database by scanning and saving, I wrote to various genetic scientists, physicians, and researchers in the field of evolutionary biology and genetics for their opinions regarding the application and use of DNA testing for family historians and genealogists, even for people who want to track and record their own lineages, family trees not only by surname, but by DNA to find out what they can.

Some people are puzzled when "and wife" is listed instead of a female and her maiden name on documents.And with women for hundreds of years taking their father's or husband's surname, doors can open to researching female lines when mtDNA is tested.

Ancient ancestry in female lineages may be traced somewhat by mtDNA. It is inherited by women and passed on to their daughters. Y chromosomes are inherited by men and passed on to their sons. Both show us clues to ancient ancestry or ethnicity even in some small ways that show expansions and migrations of people across geographic distances for thousands of years. Mutation rates and genetic drift due to isolation of small communities show the researcher where the people had sought refuge and how they expanded in clines or gradients of genes.

Where the genes are most diversified shows researchers a clue to where the genes originated rather than where they might be today. Where the genes look alike or very close, shows the people have migrated to an area only recently in the eons of time.

What are various geneticists' and genome scientists' opinions of DNA testing for genealogy research? Richard Villems,MD, Dr. Sci, head of the Department and Professor of Evolutionary Biology at the Institute of Molecular and Cell Biology, Tartu University, Tartu, Estonia, replied to my question by email, noting that, "The answer is straightforward and short: yes, DNA 'testing' is a very powerful method for genealogy research—specifically so as far as maternally inherited mitochondrial BSA and paternally inherited Y chromosome, are concerned.

"Although the current practical use is, technically speaking, far from a possible state-of-art level in case of mtDNA, the latter is, if a full sequence of mtDNA is analyzed, a very precise tool to resolve genealogy in a phylogenetically correct way already. As far as Y chromosome is involved, it would be even more so, because Y chromosome is huge compared to mtDNA—some 60 millions of nucleotides compared to about 16,500 in mtDNA. However, realistically speaking, it would take a huge technological effort to reach a stage where a phylogenetic resolution would be "final"; we are just at the beginning of a long way.

"In theory, even a son will differ from his father, in average, in a few Y-chromosomal mutations—therefore the ideal resolution would indeed allow reconstructing the biological history of this chromosome in minute details. At present, this time is still far away because of an enormous cost of such a work. Nevertheless, the fact that we do know what is possible, one may predict that any man can calculate how exactly he is related to, say, to his contemporary PM of the country—or, say to the Secretary General of the Chinese Communist Party (if that exists at that future time anyway).

"As far as autosomal genes are involved, I am pessimistic—Mendelian segregation and recombination are probably too powerful in creating noise that such a clear-cut resolution cannot be expected—never.

"Hobbyists and professional genealogists who wish to extend their family trees by confirming a link where no conventional source records exist would be interesting in having their DNA tested. Sometimes DNA tests may be used to determine whether a person is part of a larger group of people: for example, Jews will be able to confirm they are of Cohanim lineage. DNA tests are excellent for individuals who want to perform surname-based family tree reconstruction projects."

Dr. Richard Villems also wrote me this reply, when I inquired about what ethnicities my own mtDNA might reveal, "Your motif in HVS-1 is beyond any reasonable doubt within haplogroup H. Every even slightly experienced in mtDNA researcher knows that although transition in 16356 is a good guess that a particular mtDNA belongs to U4, there are enough 16356 mutations also within H. And what does it mean that 'research showed recently' that 16356C is U4—this is a very well known fact already at least for 4-5 years! But there are exceptions one ought to know as well.

Moreover, this combination you have is well present all over Europe—plenty in Scandinavia, in Estonia as well—but also in Germany and, to make it really a pan-European—also in the Adriatic area as well as in Eastern Europe down to the slopes of the southern Urals, among Turkic-speaking Bashkirs."

I also wrote to another scientist who works with human genetics, Dr. Vincent A. Macaulay, Dept of Statistics, University of Oxford, UK, who replied similarly, "Your sequence (16189-16356-16362) is almost certainly in haplogroup H. I have several exact matches to your sequence my database which are confirmed as H using other markers in the mtDNA molecule.

"Position 00073 is in HVSII, which is not in the part of the molecule that Oxford Ancestors sequenced. I think they have confused 00073 and 16073 (which is in HVSI) in their reply to you. If you had HVSII sequenced, I would be confident that 00073 would display an "A". The 16356 mutation has happened more than once, so it does not always imply haplogroup U4.

"For your information this sequence has not been observed east of Bulgaria. In my database, there are sequence matches in UK, Spain, Portugal, Germany, Austria and Bulgaria. I hope this helps: I would suggest that you seek further clarification from Oxford Ancestors."

Another scientist in genetics, Dr. Antonio Torroni, Institute of Biochemistry, University of Urbino, Italy, also wrote to me that he found one person from Crete in his mtDNA database with my mtDNA sequences. So where did my own founding female lineage come from and which country represented my direct ancestor—or did all of those countries?

Since only a sample from each country was tested and put in the various database, I wondered whether an ancestor might have been not yet tested, not in the database, and in some geographic area not yet mentioned. The journeys for the founder types have only just begun. What geographic part on Earth, what ethnicity could I ultimately call my own down to the bones? Would I ever find out?

Dr Peter Reed has a PhD in Human Genetics from the University of Oxford and was a pioneer in the use of STR genetic markers in medical research. "To non-enthusiasts genealogy is often considered an obsession with the past," says Reed. "Yet the combination of genetics and genealogy enforces how family history is integral to what we are today."

All our ancestors have contributed to our personal genetic makeup, and by examining our own genes we are viewing the DNA of our ancestors.

"Current uses of genetics in genealogy only examine a fraction of our entire genetic makeup and have largely developed from anthropological research," Reed explains. "However, the driving force of the current 'genetic revolution' has been research into human physiology and psychology, and it is from this work that new applications for genealogy will be developed."

"In five to ten years knowledge of our entire personal genetic code will be feasible, and with this knowledge will come an ability to much better understand how our ancestors contributed to our genetic makeup.

"In a few generations from now not only will knowledge of an ancestor's full genetic code be as indispensable for family history as birth and death dates are today, but family history (both genetic and social) will be a vital instrument in personal health care."

∗   ∗   ∗   ∗

The Phenomics Revolution to the Nutritional Genomics Buzz: My Positive Experiences with DNA Testing

"DNA testing is an exquisitely precise tool for answering certain types of genealogical questions, but it is clear that this technique is, and will continue to be, a disappointment to many who see it as a way of leaping over the 'brick walls' in their conventional research," says genealogy researcher, John F. Chandler. "DNA testing is at its best in demonstrating that two people or two lineages are not related within a genealogical time frame."

"When used for the purpose of proving that two people are related, it is notoriously often misconstrued. By itself, DNA testing can only show a general relationship, not a specific one," Chandler admits. "For the future, if the field continues to grow, there is some hope that DNA will offer a realistic chance to hunt for relations by looking for exact matches, but the growth will have to be at least three orders of magnitude before this comes to pass."

I wanted to find out more about Ancestry Informative Markers (AIMs) and my own ancestry than only what my mitochondrial DNA (mtDNA) could tell. I wanted to find out the racial percentages in my genealogy search and how this information could help me write my ancient or recent past journal. Was I Asian, African, Indo-European or Native American? What other mixtures?

In what percentages? How much Roman and how much Greek? How much Portuguese am I? How much Jewish and how much Scandinavian? How much Slavic and how much Middle Eastern? What were my percentages of ancestry that accumulated over thousands of years of expansions and migrations from Ice Age refugiums? From where geographically and at which times in history did each of my ancestors contribute to my genetic markers? Could I find this out—or only a clue?

The DNA of the mitochondria is the energy generators transmitted through the egg cells, according to the New York University School of Medicine's Human Genetics Program that offers genetic analysis in various studies. See the Web site at: http://www.med.nyu.edu/genetics/jewishorigins.html. The DNA of the Y chromosome told about male ancestry for males.

Women don't carry the Y chromosome and so are tested for DNA by either their mtDNA for ancestry, or their nuclear DNA. Most DNA testing companies offer mtDNA testing for women and Y chromosome testing for men, and men also have mtDNA, and may have that tested also to learn about their female lineages as well as their male-Y-chromosome lineages.

Some DNA testing companies also offer the racial percentages DNA testing. The racial percentages such as East Asian, Indo European, African, and Native American are DNA tests of Ancestry Informative Markers (AIMs). I wanted to find out what I could about my ancient and recent ancestry that extended beyond records in city directories and other cross-reference files. I took the racial percentages test to find out I was 97% Indo-European and 3% East Asian. (I was told to take anything 5% or under with a grain of salt with the East Asian.)

Was the test only revealing back to my great great grandmother and no further back? Or did it reveal the "real me" as I was 21,000 years ago? Or was I looking at a printout of the maternal lineages that contributed to my genes 500 years ago or one generation back from my birthdate? The search for a core identity and ancestral DNA records of the deep past was on.

Of all the three DNA tests that I took to find out what I could about my unknown ancestry, the DNA test that I found most helpful was the one that looked at the percentages of the various races in my ancestral history.

I took The DNAPrint ANCESTRYbyDNA test in order to look at my personal panel of Ancestry Informative Markers (AIMs). "Our test can only indicate to what percentage a person is Native American, African, East Asian and Indo European," says Carrie Castillo, Corporate Communications, DNAPrint genomics. For further information, contact, DNAPrint genomics, Inc. 900 Cocoanut Ave, Sarasota, FL 34236.

The test uses markers that have been characterized in a large number of well-defined population samples. These markers are selected on the bases of showing substantial differences in frequency between population groups and, as such, can tell about the origins of a particular person whose ancestry is unknown.

After the analysis of these Ancestry Informative Markers (AIMs), in a sample of a person's DNA, the probability that a person is derived from any of the parental populations and any of the possible mixes of parental populations is calculated.

The population (or combination of populations) where the likelihood is the highest is then taken to be the best estimate of the ancestral proportions of the person. Confidence intervals on these point estimates of ancestral proportions are also being calculated.

For example, the Duffy Null allele (FY*0) is very common (approaching fixation or an allele frequency of 100%) in all sub-Saharan African populations and is not found outside of Africa. So a person with this allele is very likely to have some level of African ancestry.

Knowing the percentage of your races may be one consideration when planning for personalized medicine. How many know the entire history of one's own ancestry? If your parents and grandparents had genetic-related degenerative diseases that are high in certain populations, would you want to know whether you carried genes from a particular ethnic group? I wanted answers to questions such as these because I'm a freelance writer and write novels about DNA.

Ten generations is roughly 250 years and within the time frame of genealogical interest, especially when we are considering the settlement of North America, because they only look at two (2) chromosomes. Ychromosomal analysis and mtDNA analysis each could only provide information on a very small proportion of a person's ancestors.

I went for my third and last DNA test to the AncestryByDNA test because it relies on sequences throughout one's genome. So the results I received from AncestryByDNA said more about a greater number of ancestors.

FamilyTree DNA also offers AncestryBy DNA's racial percentages test along with their mtDNA and Y-chromosome tests. My goal was to find out what I could for what I could afford, about my ancient ancestry through DNA testing. Perhaps it could tell me something more when oral and written records weren't within reach. Could these tests tell me anything other than that I resemble millions of people with the same mtDNA sequences, or if I were male, Y-chromosome markers? Could it provide a DNA match with someone who possibly shared a common ancestor with me hundreds of years ago? Who was my most recent common ancestor, and would that person show up in a database? The answer is only if that person took the test and asked for his or her email to be made public for contact with another DNA match.

After all, my entire genome wasn't being tested. That's not affordable on my budget. And my entire genome would reveal more personal information that the mtDNA, even the high and low resolution HVS-1 and HVS-2 sites. I still matched with millions of people. Only in my case, my mtDNA sequences are found only at a low frequency in Europe. What caused their extinctions? Or did they have more sons than daughters?

The places where my recent relatives lived were mostly unknown to me, and all I could go by was looks as perceived by others based on stereotypes based on cartoons and caricatures. Online database tables told me someone with my matrilineal DNA (mtDNA) lived in Crete. It was as good a starting point as any. So I went out and bought feta cheese for lunch melted on pizza dough.

Then one to four people per thousand with my exact sequences of mtDNA also lived in Scotland and Norway and Siena, Italy (Tuscany) and Turkey and Bulgaria, and Iceland and Austria. What part of the world did my ancestors belong to for any length of time? What would it mean to me other than an ethnic costume for Halloween or a trip to an ethnic restaurant or listening to music of that geographic place?

The first table I researched contained my HVS-1, low resolution sequences of mtDNA. Yet when I had my mtDNA tested for HVS-2, high resolution, different geographical places turned up—the Orkney Islands off the coast of Scotland, France, England, Bulgaria, and Turkey.

Which is the real me—Bulgaria and Turkey or Scotland and Orkney Islands? Grandma had red hair, but that was the paternal side. And red hair is found in all those countries. My hair's dark brown. That could mean anyplace in the world.

What could DNA testing tell me that looks did not? My social experiences were based on how others saw me. For example, at least a dozen times in my life I was stopped and "told" by total strangers what ethnic group I come from and then asked "why did I deviate" from it. When I was twelve, a lady standing next to my mother in a crowded Brooklyn bus pointed to me and shouted, "Why is that Jewish girl wearing a cross?"

Why did that lady choose that particular ethnic group based on socalled stereotypical Ashkenazi "Jewish" looks rather than Greek, Armenian, or Lebanese stereotypical looks? Maybe it was because there were more Ashkenazi Jews in that place than there were Greeks and Armenians or Lebanese? How come my mtDNA, representing two or three percent of my enriched genetic heritage from my ancient maternal line shows up Norway, Iceland, Scotland, England and other northern European countries? My face and milk chocolate brown hair looks as if it originated along the Adriatic. Was that the place?

My Roots for Real printout notes that my mtDNA sequences originated around Bar Sur Aube, France in the last 10,000 years. Scientists from whom I ask for feed back say the actual name of the geographic town of possible origin is unscientific, but you can't rule it out, either.

Was the way my face had been perceived by the public based on visual stereotypes of 'Armenoids' or 'Assyroids' in caricature or photos as portrayed in various media? Or was it someone's own ethnic group stereotypes projected on me? What about the man who saw my face as the enemy and already told me what ethnic group I belong to before he beat me up in a public train? He didn't even think of asking first.

How can you tell one ethnic group from another in a crowd or in a train going from one town in New Jersey to the next? Why did my face automatically incite anger or agitation in some people without me doing much other than walking by in front of people or sitting next to them? It wasn't based on color.My freckled face doesn't even tan.

Then what was it? The convexity of my nose? My brown hair? Freckles? Skin that don't tan? Hazel eyes? How can you tell someone from one country or one religion from another? Can you really tell an Armenian or Greek islands gal from a Polish or Romanian Jew? Does it matter? What would it incite you to kick me in the spine? Or to ask me to my face why I'm trying to "pass" as whatever blends in with the crowd?

As a member (of one high IQ society) from 1978 to 2001, I found myself sitting at a luncheon table when two members, one man and one woman, was so agitated by my name not looking like my face, that I had to show them my ID card to change the subject. They automatically assumed by my face I was Jewish. (Both of them were.) Did they only want to connect with a familiar, friendly face with whom they could feel comfortable?

When they saw the name didn't look like the face, they were astonished. But why was this reaction verbal and necessary to tell me in public at a luncheon gathering. One woman said, "What's a nice Jewish girl like you doing married to Mr. (non-Jewish)? A gentleman next to her said, "I can't believe that's your name.You sure don't look Scandinavian. You look Jewish."

I couldn't believe at a club meeting in a public restaurant that the conversation would focus on looking at my name on my ID card picture and commenting on my religion or ethnicity. What brought that up under the umbrella of making general conversation to break the ice? It was the sound of my last name. To them, it didn't look like my face. I still don't get it. What's different about my face? Then the conversation turned to rest of the group discussion the Druid religion.

Perhaps the two Jewish members felt a certain need to connect to me among a table of non-Jews. At that time in the seventies, there was less diversity in that particular California city than there is today. When I returned to another branch of the same club 24 years later, in a different city, the conversation was very different—the stock market. But back then, the conversation ran like this. From a retired dentist kidding around at a club luncheon: "Can't you talk without moving your hands? Try putting your hands in your pockets."

Me: "No. I was reared in an urban neighborhood of Greeks, Southern Italians, and Armenians—maybe. And I get nervous when I keep my hands in my pockets." Then I thought. I won't wave my hands. It just raises my blood pressure if I gesture as I talk because that puts emotion behind the words, and it's not great for my health as a genetic introvert with the high anxiety gene.

Years later, when I was in my early sixties, a woman at a bus bench in San Diego said, "You are Jewish and you come from Brooklyn." Both these people were strangers. Yet by the shape of my nose and facial features, they automatically assumed I was an ethnic group whose stereotype caricatures sufficiently motivated them to speak out to the stranger next to them.

When I was thirty and sitting in the student lounge room a blonde woman next to me reading a bible said, "Gee, you look Jewish, but you're so fair."What made these people so openly verbal to a stranger sitting next to them? Comments focused on something about the shape of my nose, dark hair, and my Assyroid skull shape.

So what was in my genes that made people stir that way and comment, or as you'll read on, beat me up for "looking like a particular ethnic group." Oddly, no one asked me what I actually was. They told me, and then proceeded to "act out." My genetic DNA tests that I hoped would reveal my actual identity or mixtures of ethnicities told a new story. That's one of the reasons why I had my DNA tested three times.

Another reason that I wanted to know more about DNA testing for recent and/or ancient origins or at least percentages of the races in my ethnic past beyond what haplogroup in Europe my founding female ancestor camped with (Haplogroup H) 21,000 years ago, was so I could write better DNA Detective or historical novels, how-to books, or books on foods.

Characters in my novels have genetics-related careers, and they solve forensic DNA problems using clues that I needed to find by reading about molecular genetics and anthropology. My characters solve important problems using DNA clues to answer questions facing DNA researchers.

They work in bioinformatics and solve cold cases using cutting-edge technology, such as in my latest forensic DNA researcher novel, The DNA Detectives—Working Against Time, published in paperback by Mystery and Suspense Press, iuniverse, at: http://www.iuniverse.com/bookstore/book_detail.asp?isbn=0%2 D595%2D25339%2D3. Or click on: http://www.iuniverse.com and click on the Bookstore, then look up my book by title or my name.

Being a DNA autodidact became my full-time hobby in the last two years. This year I wrote The DNA Detectives—Working Against Time, published in paperback by Mystery and Suspense Press, http://www.iuniverse.com. The search for identity is part of my INFP personality (taking in information and making decisions using my extroverted intuition and introverted feeling) on the MBTI TM.

I spent the decade from 1990 to 2000 exploring my personality type with the Myers-Briggs Type Indicator.Now I needed to go beyond looking at my person-ality type and which career fit best with my artistic, creative-expression personal-ity and interest in learning science as a senior citizen by reading books at home. Writing diaries as novels had evolved into writing scientific thrillers with romance, mystery, adventure, and suspense.

In the new millennium, with the media popularity of the human genome project, my search for identity through creative writing turned to DNA testing. So reading about DNA remained as my primary hobby.

I'm over 61 and have little to do all day but hang around senior centers and go to genealogy club meetings. Reading about the latest in DNA research took me light years beyond genealogy. Of what use is a surname when the name was con-stantly changed every other generation?

Being a home-based non-driver with little experience in anything more than making use of my Masters degree in English/creative writing and minor in anthropology and illustration by spending the last four decades writing several books a year, with 29 recently published, I put up a Web site at http://dnadetec-tives.tripod.com. I had two choices on how to spend this decade—knitting at the senior center or writing novels about how people use knowledge of DNA.

My first experience with DNA testing was to have my mtDNA tested to find out anything I could about my ancestry which had been more or less unknown to me, other than having brown hair, hazel eyes, fair skin with a yellow undertone, and freckles. Back in 1964 I was beaten up in a train going from Asbury Park to New York for looking like a particular ethnic group.

Since then I sometimes wondered what is there about my particular facial fea-tures that drove someone to beat the heck out of me when I was pregnant and in my twenties. The epithets and body language conveyed to me that I was "crushed" both for looking like a group some people like to insult or strike and for disobeying the passenger's order to wait until the train stopped, even though passing in front of him wouldn't have changed his life in any way.

I wasn't prepared for hate. Someone with fair skin and freckles, green eyes, wasn't told to watch for it. The color ofmy black wig and the shape of one or two of my facial features set off the rage of epithets hurled at me. We were both "equal" passengers in a train, and I was showing courtesy by standing at least a few feet from him as I passed in front of him between the cars to return from the restroom to my seat among my family. This wasn't my fault, and I shouldn't have felt that if only I wore a different color wig, I would not have been noticed.

That day I wore my black instead of my blonde wig. I was twenty-two and not really aware of hate crimes or how the color of a wig or the shape of a facial feature might set off a passenger in a commuter train.

Back in 1964 I was riding in a train going from Asbury Park, New Jersey to New York. Out of the blue a middle-aged passenger began tossing loud ethnic epithets at me and demanded that I shouldn't pass between the cars where the restroom was located and the car where my seat was located next to my relatives.

When I ignored his demand to wait until the train stopped in Elizabeth, about a half hour away and tried to pass in front of him to return to my own seat, he grabbed my head, crushing it painfully between his knee and the metal train car door, yelling more ethnic epithets, and kicking me with all his might at the base of the spine. No one came to help. Gee, what if I had been deaf and didn't hear his demand and just went back to my seat because that's what one does after using the john?

Only he called me a name mentioning an ethnic group. It reminded me of the plastic surgeon I met in public when I was fifteen who commented, "Let me take that hump off your nose." I was thunderstruck the way the plastic surgeon talked me me at fifteen. I replied, "I'm an American. No foreign language was ever spoken in my home, and my parents were born right here in the heart of Brooklyn's Little Italy. My religion or no religion is private as is my ancestry. And if I had an ethnic core identity or another language, all the richer for my entire tapestry." For now that core identity remains the human race.

Curiously, I would have loved to find out my ancient deep ancestry. Years later, the DNA tests told me where my single female ancestor lived 21,000 years ago. I wish I could have found out where all the mixtures on the paternal side and maternal side lived in those past millenniums. DNA testing of the entire genome someday might give me more clues for my map of expansions time capsule. Now, I can even tailor my food to my genetic expression.

In any case, that visual familiarity weighed on my mind enough to stir me to find out just how many races or ethnicities I had in my ancestry and why the site of dark hair and Middle Eastern (Neolithic farmer) features would arouse such explosive anger in a European-looking stranger that I had never spoken to, met, touched, or seen, except for my desire to walk past him to a second car and return to my seat from the restroom.

He couldn't have known that I was three months pregnant, married, and American. Not that my ethnicity mattered to anyone on a train, but out of curiosity I wanted to find out where my ancestors had been for the past fifty thousand years.

I've written an article on that travel piece. But getting back to DNA, I saw an announcement on the Web about Oxford Ancestors from http://www.OxfordAncestors.com and had my mtDNA tested.

That's the matrilineal line of ancestry for women. A little cytobrush arrived in the mail. I rubbed the inside of my cheek, and mailed it across the Atlantic to Oxford, England, and two months later, a pretty chart came back with a page on how the mtDNA (mitochondrial DNA) testing was done, a pretty chart linking my mtDNA clan of Helena (or H haplogroup) to everyone else's' in Europe. (A year later the chart changed to how everyone in the world is linked by mtDNA). I also received a printout of the letters of my mtDNA.

The letters were a printout of CGAT, the letters of everyone's DNA, and the chart didn't have the letter of my Haplogroup, but instead, a name given by Oxford Ancestors to the first female ancestor of the H haplogroup clan in Europe,Helena. There was no mention on the chart that Helena was a moniker for the H haplogroup which is one of the most common in Europe, making up about 47% or more of Europeans and 6% of Middle Easterners.

The only clue to my sequences were three little red letters—C showing how I differed in three places from the Cambridge Reference Sequence (CRS) by transitions or mutations of nucleotides, which I was compared against. I still wanted more—to find what ethnic groups had larger numbers of the same sequences of mtDNA as me. Also, I wanted to know where the most variation occurred, in what part of the world.

Perhaps that would give me a clue to the origins. It's not so much how many people today live in an area with your sequences, but how much variation there is that tells you how old your mtDNA is. So I was looking for a founder female, a single person or a coalescence point. For H haplogroup, the coalescence point in Europe had been about 21,000 years.

Now I asked whether H arose from a common ancestor? If so, where did the ancestor come from and was she also an H, or an HV—since H and V split off from HV and HV evolved from pre-HV—somewhere and sometime before 28,000 years ago. H had arose in the Pyrenees 20,000 years ago, but H had a common ancestor elsewhere.

Even HV had a common ancestor, pre-HV, somewhere in the Middle East, further back it time. Or did H arise by itself as a mutation in Europe during the height of the Ice Age 21,000 years ago between the coast of SW France and N. Spain's Pyrenees?

The printout of letters didn't mean much to me then. I needed sequences for mutations written in numbers. I found the sequences numbers by asking at one of the genealogy and DNA mailing lists on the Web whether anyone could tell the sequences in numbers from my printed out letters. All I had were the three mutations in red—three "C" letters.

So to answer in part these questions, I wrote to an acquaintance at the Whitehead Institute at MIT (genomics research division) who told me that the three "C" letters (mutations) could be put on another table that I never received or saw before. This table now had numbered sequences.

The table was available to anyone on the Web. I emailed the acquaintance my low-resolution sequence numbers of the HVS-1. Now I had at last sequences to compare so I could look up what countries of the world these appeared in presently, even though there was no way to tell whether my particular ancestors lived in any particular location at any time.

I looked on the Web at Victor Macaulay's HVS-1 tables at: ttp://www.stats.ox.ac.uk/~macaulay/founder2000/tableA.html and found that my sequences H haplogroup mtDNA of 16189C, 16356C, and 16362C all showed a mutation from T to C and were found with that transition on the table to appear in Spain, England, Austria, and Bulgaria.

Any place else? I emailed an mtDNA research team member, Kristiina Tambets, Estonian Biocentre and Department of Evolutionary Biology, Tartu University, Tartu, Estonia, who took part in a research study appearing in an article that I read in *Archaeogenetics*, a book published by The McDonald Institute Monographs, edited by Colin Renfrew and Katie Boyle, and Tambets emailed me several sequences, including mine, from her unpublished database.

The few sequences she sent me from her unpublished database classified my sequences as H2b, but she emphasized that I should be cautious as the material was not published. And my sequences in her database of thousands also included other countries listed under H2b, which included my sequences, (a division of haplogroup H).

The sequences listed as well countries that were found to contain at least one or more persons with my HVS-1 sequences, such as: Bashkortostan, Turkey, Crete, Croatia, Albania, Hungary, Portugal, Germany, Iceland, Spain, Norway, Sweden, Komi (Finland group) including the other countries—Spain, Portugal, England (UK), Germany, Austria, and Bulgaria as well as the four areas listed on the Web on Macaulay's tables—England, Spain, Austria, and Bulgaria for H mtDNA haplogroup, HVS-1 region, with transitions at 16189C, 16356C, 16362C. Family Tree DNA also found another transition—16519C as they tested more than 400 base pairs. Oxford Ancestors tested 400 base pairs. I eventually also had my HVS-2 high resolution mtDNA area tested.

The number of samples obtained from each country also was listed in each of the database sequences or tables that I studied. Victor Macaulay's mtDNA (mitochondrial DNA) tables also are on the Web at: http://www.stats.ox.ac.uk/~macaulay/founder2000/tableA.html.

My acquaintance from MIT whom I met from Internet correspondence on a mailing list of interest to people working with DNA, wrote that I was probably close to Bulgar/Turk or Karelian or a mixture of all three, when he looked at my sequences back in 2001. He didn't say how he came to that conclusion. However, he asked me to look at all my mtDNA sequences when searching the tables online.

I assume that the sequences might have shown up among these ethnic groups. So I didn't really learn the process of how he found it out.

My research showed England, Spain, Austria, and Bulgaria on Macaulay's tables online. So I needed to learn much more about how to look up sequences on tables that are on the Web for all to access.

If I removed the 189, a fast-mutating site, and only looked at 356 and 362, then it fit well into Karelia, Bulgaria, or Turkey. The 356 mutation, fit with Armenia. But what if I looked at all my sequences? Then other countries were on the tables.

In 2001, I received a certificate from Oxford Ancestors saying I was a Helena (haplogroup H). I thought this was awesome, because I was reared thinking I was in ancient times, at least in part Middle Eastern or from the Caucasus mountains, where U haplogroup is common and H is found at lesser frequencies than in Europe.

In 2002 I received a certificate from Oxford Ancestors saying I was an Ulrike (haplogroup U4). I wrote Vincent Macaulay asking how can one tell which is which with the same mutations? From his tables and email, he said to find out whether I had a G at position 00073 at HVS-2.

He emailed me a note saying that chances are almost certain that I had an A at that position 00073 on HVS-2, if I'd look. Certainly the three mutations I had were in his online database tables.An A at position 00073 of HVS-2 would make me H haplogroup. A G at position 00073 would make me a U4, but the only way to be sure is to do a high resolution test of the mtDNA for HVS-2.

So I wrote to Dr. Villems from Estonia. My sequences are also found in that database. My sequences showed up in several North Eastern and North Central European countries not mentioned in any of the English databases or on the sequence matches sent to me from one of the DNA firms. Some countries overlapped such as Bulgaria, Spain, Portugal, England, and Scotland. The Eastern European database also included my mtDNA sequences in Bashkortostan, Siena, Italy, Crete, the Ukraine, Germany, Poland, and along the Adriatic. With my mtDNA representing only 2 to 3 percent of my genome, how far back would the ancestry go? People in those countries may not be relatives in the modern sense, but they all at one time had the same female ancestor—one woman with those sequences. For me that was enough to make them siblings of a sort.

I looked at my sequences that came from Bryan Sykes, from Oxford Ancestors. Bryan Sykes, MA PhD DSc, is Professor of Human Genetics, University of Oxford, and with Oxford Ancestors. He has written several books on the history and geography of human genes, including The Seven Daughters of Eve and a book on the sons of Adam, genetically speaking. Anyway, Villems also agreed I'm probably an H, not a U4 (what Oxford Ancestors named Ulrike). I had also written to Vincent Macaulay since his mtDNA tables are on the Web where I could look up my mtDNA sequences and see what countries or areas they appeared in.

On Macaulay's tables, my mtDNA sequences showed up at a low rate in Spain, England, Austria, and Bulgaria. Scientists use mtDNA and Y-chromosomes for ancestry. Again, it represents only two to three percent of my entire genome, which would include genes recombined each generation that I received from both my parents and other ancestors in varying percentages.

The question came up because one of my sequences, 16356C also is found in mtDNA haplogroup U4. Dr. Sykes and I corresponded by email several times, and Dr. Sykes is most helpful and emailed me on January 15, 2002. He wrote, "David Ashworth from Oxford Ancestors has shown me you message and the replies you received from Drs. Macaulay and Villems about whether your DNA sequence places you in the clade of U4 Ulrike or H Helena.

"David tells me that on your original certificate, issued in August 2000, you were placed in the clan of Helena but that when you were sent a replacement you had become a daughter of Ulrike instead. Of course your actual DNA sequence hadn't changed, but the assignment of you clan had.

"It may help if I explain how that is done. Clans are defined by a mixture of two sorts of genetic markers, the variants in the control region sequence and the variants at a number of other sites around the mtDNA molecule now generally called SNPs (short for Single Nucleotide Polymorphisms). These are usually designated as +4643Rsa1 or+11329Alu1 etc as you have pointed out in your messages.

"Vincent's classification on the website and in the papers you refer to contain a complete list of these variants. What they mean is that a restriction enzyme recognises the variation and either cuts or does not cut the DNA at that point. Since your hobby is reading about DNA, I am sure I don't need to explain what a restriction enzyme is. So, take+4643Rsa1. That means the enzyme Rsa1 cuts at base number 4643. The variant-4643Rsa1 means that the enzyme does not cut the DNA.

"Vincent and I, with Martin Richards, spent a great deal of time correlating the control region variants with the SNPs by analysing both on several hundred (it may even have been thousands) of mtDNAs and this was an important part of distinguishing the different clans. On the whole these types of variant are more stable than some of the control region sequence variants but not always and ideally every DNA should be tested for both. However, that would put the price to customers up hugely because each one of the SNPs had to be done separately—although I know that Oxford Ancestors are looking into offering this service.

"But even that would not guarantee completely accurate assignment in every single case. Only sequencing the entire mtDNA circle of sixteen and a half thousand bases at astronomical cost would do that—though even that would not be any good unless you had at your fingertips a database of thousands of other complete sequences with which to compare it and only a handful have been completely sequenced to date.

Also, as more work is reported, the evolutionary networks will change. "What I am getting at is that no system is foolproof. The Oxford Ancestors service, to keep it affordable, only sequences the control region. Then the sequence is compared to a database which holds other sequences which have been examined for SNP variants. If the customers control sequence matches up with one of these then it is assigned to the same clan.

In other cases, where there is not an exact match, the database is searched for close matches or sites which are characteristic of particular clans. In the case of your sequence which has variants from the reference sequence at 189, 356, 362 (we delete the 16 prefix for HVS1) two of the three variants are quite unstable—that means they can mutate back and forth. Position 189 is one of the least stable of all and position 362 is not very far behind. Position 356 is far more stable and is also characteristic of clade U4, whose clan mother is Ulrike. However, it is not completely stable and does crop up in other clans—one of which is Helena.

"So the sequence 189, 356, 362 could be in the clan of Ulrike mutating at the unstable positions 189 and 362 away from the U4 root sequence of 356. Or it could be in the clan of Helena with a rare variant at 356 taking it away from 189, 362.

One way of telling the two apart is to look at the variant at 073. This is actually in HVSII and not HVSI and that was a source of confusion in some of the email exchanges I have read.

"Oxford Ancestors doesn't do the 073 test, as you know, so the sequence was assigned on the balance of probabilities to Ulrike. I have now had a chance to compare the sequence to some new research data of my own from Britain in which we did do the 073 test and found five exact matches which carry A at 073, indicative of clade H. So I think you are probably correct and are indeed a daughter of Helena rather than Ulrike. This means that the mutation at 356 would have occurred on a Helena background rather than the 189 and 362 variants occurring on an Ulrike background.

"That might explain why you were originally issued with a Helena certificate in August 2000. At that time, the service was being sent out from my laboratory before Oxford Ancestors acquired its own premises. That means that whoever did that first assignment, and it may well have been me, did recognize the ambiguous nature of the 356 mutation in that particular sequence but that piece of information was not properly transferred to the new set-up—and that is my fault.

"I must thank you for clarifying the assignment of this particular sequence. It is a changing field and your observation has helped it move on one more stage further. I am sure Dr Ashworth will want to issue a new Helena certificate. And of course, I hope you are pleased to have moved back to your original clan."

*   *   *   *

Belonging to Helena's clan sounds more personal that what it means—a member of haplogroup H, presumably one of the Seven Daughters of Eve, the founding "clans" or haplogroups found in Europe in prehistoric times. But even H haplogroup had to have a common ancestor before heading for Europe in the middle of the Ice Age 21,000 years ago. And that ancestor had to come from presumably somewhere in the Middle East.

Haplogroup H is the most frequent cluster in the Middle East and the Caucasus. It's present at a frequency 25 percent. In Europe, it's found much more frequently at around 46 percent. There are more haplogroup H people in Europe than there are in any other part of the world.

The mtDNA haplogroup H represents a female ancestor who lived in the Middle East up to 28,400 years ago. Some surviving daughters then went to Europe, and some stayed in the Middle East. However, the European daughters' mtDNA with Haplogroup H have an "age estimate" of only up to 21,400 years. So the arrival of or mutation to H in Europe began more recently in Europe than it did in the Middle East. Yet today there are more Europeans with haplogroup H than there are in the Middle East with the same matrilineal lines of mtDNA haplogroup H.

Other mtDNA haplogroups of female lineages are much older, if we look at mtDNA coalescence times and founder effects. For example, mtDNA haplogroups T and J both date to around 50,000 years before the present in the Middle East, but more recent dates in Europe. Haplogroup J is found today at its highest frequency in Arabia, making up 25 percent of the Bedouin and Yemeni population.

It's also found in Europe, also a more recent arrival. Perhaps mtDNA haplogroup J entered Europe around 10,000 years ago or less, probably with the agricultural revolution of the Neolithic age, after the last Ice Age ended around 12,500 years ago.

The oldest mtDNA in Europe is the female lineage of U mtDNA haplogroup. It's around 50,000 years old in the Middle East, perhaps around 45,000 years ago in Europe, and from U comes many sub divisions such as European-specific U5, North African U6, and Indian U2i, all showing an origin of around 50,000 years ago. The first female lineage out of East Africa into the Middle East was L3, but mtDNA haplogroup U also followed out of Northern Africa and went to Europe as well as most of the Middle East, where it's found today.

An interesting theory is that 50,000 years ago, the route from Northern Africa was closed until 44,000 years ago, so people coming from Asia or Africa into Europe through the Middle East had to come from India across the southern route until the route from the Middle East opened up to get to Europe, and that was around 44,000 years ago.

At that time the Fertile Crescent around what is today Iraq opened up as well as the Levant. By then, people could use a northern route across the Bosphorus from Turkey to Greece and enter Europe. Those in what is today Greece then would have gotten trapped in the Balkans by 25,000 years ago, unless they made it to the refugiums in southwestern France and northwestern Spain near the Mediterranean and at the foot of the Pyrenees. Some made it.

That's why we have the wonderful cave paintings of large animals in southwestern France, particularly the Dordogne valley. Art bloomed when the Ice Age was raging at its maximum and the open tundra or fields of grass was a steppe for animals and hunters. As soon as the Ice Age ended and forests took over most of Europe, the wonderful art subsided into a dark age that began around 12,000 years ago. Then the early farmers came up from the Middle East.

They now make up about 20 percent of Europe's population, and the idea of farming caught on during the next few thousand years. There was an agricultural frontier that separated hunters from farmers, and it lasted thousands of years.

What puzzled me is why today H haplogroup is nearly missing from the Arabian Peninsula. What made it leave, the changing climate or low population frequency? For more information I turned to an excellent article titled "Tracing European Founder mtDNAs" by Richards, et al published in the American Journal of Human Genetics, 67:1251-1276, 2000. I read about mtDNA haplogroup V, which appears to have expanded in Europe around 13,000 years ago. Then I re-read Bryan Sykes's book, The Seven Daughters of Eve. It was time to take the next step.

So I went to Family Tree DNA for more information. I had my HVS-2 (high resolution) mtDNA tested to find out more about what my ancestry might be and whether I was a U4 or an H haplogroup. The two single female ancestors who founded H and U4 in prehistoric times came from different parts of Eurasia.

Family Tree DNA sent two small felt-tipped serrated pads with a push stick that pushes the felt pad from the holder into a small vial of preservative solution. I brushed the inside of my cheek several times, put the two pads into the two tiny vials of solution, and sent it back. I had a little difficulty poking the stick to drop the pads, and in the end had to pull the pads firmly from the sticks and drop them into the vials with my fingers. (I washed my hands first.)

Family Tree DNA tested more base pairs. I found that indeed there was an A at position 00073 in HVS-2, just like Victor Macaulay told me I'd probably find if I looked. That made me a member of H mtDNA haplogroup. Family Tree DNA also found a new transition to the letter C at position 16519. Family Tree DNA tested 540 base pairs as compared to the 400 tested at Oxford Ancestors, which is close to the minimum that needs to be tested to find out one's mtDNA haplogroup results.

Bennett Greenspan of Family Tree DNA emailed me an answer to my inquiry of what question he would like to see answered in a book on genealogy through DNA testing. "It would be to explain the direct line of descent…male to male to male, with no mixing of the genders. That's what people seem to have a hard time with."

What I learned from email correspondence from Family Tree DNA is that my mtDNA won't tell me when the last mutation on a string happened, but H haplogroup in its form as H haplogroup appears on the CRS (Cambridge Reference Sequence) without all the mutations that I have occurring over thousands of years of evolution, is a little over 20,000 years old. (I visualized H as the sequence of the Cro-Magnons of the Mesolithic caves of SW France and Northern Spain.)

Could my ancestors have painted that fine art work on caves at Lascaux or Altamira during the height of the Ice Age? Since then my mtDNA has mutated away from the Cambridge Reference Sequence (CRS), and perhaps the last mutation that occurred in my HVR-sequence happened 2-10,000 years ago. Should I look to Europe or the Levant for my founding matrilineal line?

And I wish there was a way to also see what ethnicities my patrilineal line also contributed. I don't have the Y chromosome to trace the Adams in my family, but can a test of percentages of races give me any clue? What tests were within my budget?

Therefore every person I find that matches me exactly on HVR-1 is a nice clue because I am closer to them then to any one else on the direct female side. I like the way Family Tree signs "email me anytime" at the end of each email address. Response time was fast, and my questions were answered.

So I'm thankful for all this information I found useful and practical. They have databases, search engines, and surname projects. Yet I still needed to find out more about what races are in my ancestry. I moved on and found another DNA testing company that even has a chat room.

There are about 3 billion DNA letters (technical name: "base pairs") in human DNA. You can picture these letters as beads on a keychain, each one labeled with an A, C, G, or T. The letters stand for adenine, cytidine, guanine, and thymidine, always referred to as A,C,G, and T.

These nucleotide bases are joined together, one after the other like a molecular string of pearls called DNA (deoxyribonucleic acid).

Oxford Ancestors tested only 400 of my HVS-1 sequences. Family Tree DNA tested 540 and found a new mutation or transition16519C in HVS-1. Then when I had HVS-2 tested at Family Tree DNA, I found the transitions also were very common on the CRS and were 309.1C, 315.1C, and deletions at 523-and 524-. Any mutations I had that showed up in my HVS-2 results were more common in the population, and the rare sequence was on the CRS.

I looked up HVS-2 sequences on Victor Macaulay's tables on the Web and found additional countries such as Orkney Islands off the coast of Scotland, Norway, Turkey, Bulgaria, Scotland, and England matching my HVS-2 results. So far, I had no clue as to whether my maternal linewas European or Middle Eastern or both. You have to look at HVS-1 and HVS-2 together and not look at them separately when you are trying to find out your nearest common ancestor.

Since my family is so multicultural, I wanted to find out what the DNA said. Family Tree DNA had a database and a name search, but I wasn't able to get into the database or understand what I had to do to check for matches with similar names, but I was very happy with the results.

Now I had low and high resolution mtDNA. That still wasn't enough. There was no way I could afford to have the entire genome checked. Some Web sites say they will check the whole genome for ancestry or other for a fee that I couldn't afford. Some people want their whole genome done for medical reasons or to make sure medicine is matched to their personal genome.

In a few years, perhaps anyone can have his or her nuclear DNA and entire genome checked for ancestry or to match the right diets or medicines to individual genetic situations. Right now, a whole genome check is priced in the thousands at some sites and at varying prices too high for me.

Also, what I learned from Family Tree DNA, that I never found out from Oxford Ancestors, is that when I had the HVR-2 tested, Family Tree had included some very fast moving, potentially unstable markers to attempt to break down the time when I find a match on HVR-1.

This new information helped me move forward in my research for my own ancestry through DNA, something I heard couldn't be done yet, (other than in the distant past.) Yet I was getting closer. Now it was time to move on to the percentages of races that are found in my DNA.

That's when I turned to a BioGeographical Ancestry test from Ancestry-ByDNA. It's a company that recently presented a complex genetics classifier for personalized medicine. AncestryByDNA's CEO is Dr. Tony Frudakis, of DNAPrint Genomics, Inc. 900 Cocoanut Ave, Sarasota, FL 34236. The Web site is at http://www.dnaprint.com/.

The Web site for testing ancestry such as percentages of various races is located at: www.ancestrybydna.com. They also have a chat room.

DNA testing is about personalized medicine as well as finding out about ancestry and other genetic-related questions. It's also about an informatics platform for genotype pattern recognition.

Here's your gateway to understanding the genetic basis for complex trait determination. DNA testing is about looking at pioneers in the emerging field of post-human genome phenomics. By putting the genomics puzzle together, you become part of the phenomics revolution with ancestry testing or other genotyping services. It's about research on individual genetic responses to certain medications as much as about finding your ancient or racial percentage ancestry.

From the DNA kit I was sent with printed information, I mailed in a DNA sample obtained from rubbing a soft piece of felt inside my cheek on a felt-tipped, serrated swab, which I air-dried and placed in an addressed return envelope and mailed back to DNA Print genomics and asked to be tested for percentages of various racial groups such as African, East Asian, European, and Native American.

Some post offices don't want you to mail biological material, such as DNA. Most post offices, though allow you to put the envelope in a regular mail box, since it only required two first-class stamps to return, weighing around two ounces or less. The envelope contained a signed form with my address so my printed out results could be returned.

You can also get your results by email. To be safe, I placed the return envelope with the two swabs inside of another mailing envelope in case a mailing machine cut a hole in the self-addressed return envelope containing my two pieces of felt with my DNA on it from inside my cheek.

If you air-dry the DNA on the small felt tip before you put it in the envelope, you won't get saliva wetting inside of the mailing envelope. When checking for racial percentages from each parent, the DNA testing goes beyond only testing the mtDNA for women and Y chromo-some and/or mtDNA for men.

Those results are limited to only a very small number of your ancestors in your ancient past, such as your female founders for mtDNA (women) or Y chromo-some and mtDNA for my ancestors who lived perhaps in Ice Age refugiums thousands of years ago.

The test for racial percentages relies on sequences throughout your genome, so a DNA testing company can say more about a greater number of ones ancestors. That's why I went to the BioGeographical Ancestry (BGA) test. More companies are offering it, but I saw it first at ancestrybydna.com. BioGeographical Ancestry (BGA) is the term given to the biological or genetic component of race.

I was looking for what the BGA offered—a simple and objective description of my ancestral origins in terms of the major population groups such as: Native American, East Asian, Indo-European, and sub-Saharan African. BGA estimates represent the mixed nature of all of us in current populations.

Most of us have racial mixing in the distant past that we don't know about yet. My search for the best nutrition for my genotype as well as a search for self-identity motivated me to take these types of DNA tests.

My mtDNA sequences are pan-European. At the present, only the test of racial percentages gave me some clue to where my ancestors might have come from way back when and maybe more recently, too. What's beyond? There's the nutritional genomics field to explore with genetic testing for the right food for my genotype. Then there's the phenomics revolution in healthcare customized for one's genotype. Reading about the latest research will clue me in to what comes next. It's more about what consumers are taking into their own hands, taking control of that interests me, in relation to DNA research and the whole field of molecular genetics.

The reasons why you would look at more of your genome is useful, not frightening. And all the new ways to look at your genome are becoming more personalized and relevant.With phenomics, the future moves towards personalized medicine. And DNA testing offers a more personalized genealogy for us history buffs and novelists.

Let's say you tested your mtDNA to find your female lineage, but you want to look up female relatives and don't know their maiden names. My own great grandmother's death certificate didn't list a maiden name. The next step would be to look at her marriage certificate.

I could also look at the City Directory for a particular city in 1889 to see where she lived, looking first under her father's name. I could try the US Census records for 1890 also in that city.

My grandparents were small children at that time living in a large house and farm in upstate New York, and all I know to begin my search is my great grandmother's first name and her father's last name, and I have my mtDNA, female lineage high and low resolution sequences.

Those would be the same in my great grandmother as in my present grand daughter. The mtDNA stretches back to a single female founder 21,000 years ago in Europe and further back in time, the usual cereal belt. I know my sequences are quite popular in Crete but also popular in Scotland and Norway. What this means is that my mtDNA might have been ancient enough to have spread over a wide area—from Iceand to Bashkortostan.

If the mtDNA was less ancient, it might have remained in a relatively smaller area until present times. I know the sequences mutated from H mtDNA haplotype that existed in Spain and France 21,000 years ago to my particular sequences found in a variety of countries. The sequences themselves mutated from H with no mutations to H with four mutations sometime in the last 10,000 years. So jumping to the last century, I could look at city directories to find female ancestors whose maiden names were never recorded on any certificates. My grandmother, Frida never had a maiden name listed on her death certificate. How could I find out what her maiden name was and what country might she have possibily lived in before she left Argentina for the US in the 19th century?

City Directories often list maiden names and the names of all residents living in a home, particular before homes had telephones. Where else can you look on the Web to start besides the usual beginning genealogy Web sites? You might order marriage certificates or death certificates as well as birth certificates to find maiden names.

The National Center for Health Statistics in Hyattsville, Maryland is also on the World Wide Web at: http://www.cdc.gov/nchs/howto/w2w/newyork.htm. For example, to view an address where to write to in order to purchase a copy of a marriage certificate for New York State, the Web site will help provide information as to where to write to in order to purchase copies of marriage licenses for New York, for example, from the year 1880 forward.

For other states, check the Web site at The National Center for Health Statistics under each state. You can also search foreign countries for records. You could also look at the Ellis Islands records, if a relative came to Ellis Island, and even view a picture of the ship at the Ellis Island Online Web site at: http://www.ellisisland.org/.

One good place to start is http://www.cdc.gov/nchs/products.htm for publications and products. Click on http://www.cdc.gov/nchs/nvss.htm for Vital Statistics. Click here http://www.cdc.gov/nchs/releases/96facts/mardiv.htm for marriage and divorce statistics.

CHAPTER 15

▼

GENE POWER THROUGH SMARTER FOODS AND FAMILY HISTORY

Companies That Bring the Power of DNA Technology to Your Home: The DNA Testing Companies of Interest to Family Historians and Genealogists

The Power of DNA Technology in Every Home. This is the slogan of the Gene-Tree DNA Testing Center that supplies DNA testing applications directly to the consumer. "We would like to demonstrate how we provide the science of DNA analysis applications directly to the consumer, allowing them to conduct their own research projects in the comfort of their own home," says Terrence C. Carmichael, MS, founder of GeneTree DNA Testing Center.

"By examining the Autosomal STRs, Y-chromosome STRs and mtDNA sequence analysis and RFLP, GeneTree is helping people (such as genealogists, anthropologists, and just those generally interested) uncover their deep ancestral migration patterns, establish biological relationships with relatives 1-50 generations apart, and uncover the mysteries of past and present relationships. These services are wonderful, and prove to be of great value to the consumer, whether it is for immigration purposes, assistance with genealogy or anthropology research, or for answering the simplest of questions, such as 'are you my father?'"

After receiving his MS degree, Carmichael went on to receive a Professional Designation in Marketing and Sales from UCLA. Terry has worked at the DNA laboratory bench for 4 years and spent 9 years providing Product Development, Technical Consulting, and Marketing for the DNA purification industry, working for companies such as Bio-Rad and QIAGEN. In 2000, Carmichael co-authored a book titled, "How to DNA Test your Family Relationships".

Having started 2 successful businesses, Terry is a visionary. He has applications submitted for 2 separate patents; one for applying DNA profiles to identification cards and the other for a new high-throughput DNA purification product held by Bio-Rad Laboratories. Below are the Web sites for some of the products offered by GeneTree DNA Testing Center.

GeneTree Products:
http://www.genetree.com/servlet/moonshine/goto?page_url=/products/productgroups.jsp
Y-Chromosome Information:
http://www.genetree.com/servlet/moonshine/goto?page_url=/products/product.jsp&id=7
mtDNA Information:
http://www.genetree.com/servlet/moonshine/goto?page_url=/products/product.jsp&id=8
Native American Assessment:
http://www.genetree.com/servlet/moonshine/goto?page_url=/products/product.jsp&id=20
Biography: Terrence C. Carmichael,MS
Terrence Carmichael, MS
(888) 404-GENE (ext. 207)
****

GeneTree DNA Testing Center
3150 Almaden Expressway, #203
San Jose, CA 95118-1253
Phone: (888) 404-GENE/(408) 723-2670
Fax: (408) 723-2671 http://www.genetree.com/
****

The Power of DNA Technology in Every Home
FamilyTreeDNA
Here's how genealogy and DNA testing interacted together for the Craycraft Surname Project from FamilyTreeDNA. The Craycraft DNA project involves the surnames: Cracraft, Craycraft, Cracroft, Craycroft, and Craecraft. For purposes

of this article, sent to me by FamilyTreeDNA, the use of the surname Craycraft refers to all the possible spellings noted above, unless otherwise indicated. The surname first appears as Cracroft, in Hogsthorpe, England in the early 1200's with one Walter de Cracroft son of Humphrey Fitz Walter.

It is later found in Hackthorn, England in the early 1600's, and today a Cracroft family still resides in Hackthorn Hall. Currently there are about 10,000 people, with this surname, and the surname is found in the following countries: England, USA, Canada, and New Zealand.

The first known emigrants to the Americas with this surname were John Cracroft and his wife Ann who emigrated from Lincolnshire, England in about 1665. The passenger list has John and Ann recorded as Creacroft or Creacraft. Today the descendents of John and Ann mostly spell their surname as Craycroft. Later, Joseph Cracraft/Cracroft came to the America's circa 1702 from Lincolnshire, England. He had 5 sons.

The objectives of the DNA project for phase I were the following: 1.Verify the documented genealogy research for the Joseph line in the US 2.Verify the documented genealogy research for the John line in the US 3. Determine whether the US Craycrafts are related to any England Cracrofts. To meet the objective of verifying documented genealogy research, multiple participants were required due to the many branches between now and the identified most distant ancestor. The Lines tested are shown below, labeled by the most distance ancestor's name.

Joseph Cracraft, born Lincolnshire, England: Descendents of 3 of his 5 sons. John Cracroft, emigrated from Lincolnshire, England in about 1665 Cracroft family residing in Lincolnshire England, descendents of Walter de Cracroft.

The results from Phase I:
1. The documented genealogy of the descendents of three sons of Joseph was verified. A search continues for descendents of the two other sons.
2. The documented genealogy of John's descendents was verified.
3. The DNA results of Joseph and John's descendents matched as well as these DNA results matched those of the present day Lincolnshire family of Cracroft.

The surprises uncovered:
1. It was rumored that a female ancestor had a male child out of wedlock, and the child assumed the family surname. This was confirmed by the descendents in this branch having different DNA.
2. An adoption or non paternity event was discovered when a descendant's DNA results did not match. This event was confirmed by testing descendents from a

branch earlier in the line, which did match the Craycraft DNA results. A Phase II of the project is planned.

```
===================================
```
Spot Light: Austin Research Validation
```
===================================
```

Validating research with DNA testing: Descendents of John Austin, JR (1726-1795)

Records in the US in the 1700's are scarce, and researching in this time period is very difficult. After 26 years of research, there was a preponderance of circumstantial evidence that showed that William Austin (b. Bet. 1750—1760, Hallifax County, Virginia or Surry County, North Carolina) was a son of John Austin, Jr. (born 14 Sep 1726, James City County, Cornwall Parish, Virginia (now Halifax & Pittsylvania).

After 26 years of exhaustive research, the researchers turned to DNA testing to find the answer. The objective of the testing was to determine whether the circumstantial evidence from research showing that William was a son of John Austin, Jr. could be proven or disproved.

To accomplish this objective, two participants were selected. One participant was a documented descendent of William Austin. The other participant was a documented descendent of one of the sons of John Austin, Jr., Isaiah. The results of the DNA test confirmed the research.

```
===================================
```
Spot Light: Roper Surname Project
```
===================================
```

Objective: Determine if any of the Roper Lines are related. There are many Roper Lines in the US and England that can not be connected by documents. For Ropers with ancestors who resided in new Kent County Virginia, the courthouse fires destroyed all records before 1864, severely limiting the ability to trace a family tree past 1864 and connect with any other Roper Line. To date, 25 participants have been tested. The geographic representation of these participants include:

18 US
2 Canada
1 Australia/England
4 UK

The surnames tested are Roper and Rooper. The results to date have exceeded expectations and been a significant contribution to Roper genealogy research. The major results include:

1. The majority of Roper Lines in the US match 24/25 or 25/25. One of these lines can trace their ancestor back to John Roper, born 1611.

2. County Norfolk in England has been identified as the ancestral homeland.

3. Most of the branches of the US Ropers have a different mutation, enabling a branch to be identified by the mutation.

In addition, research was confirmed when a participant who descended from a female Roper did not match, as would be expected. The next phase of the Project involves more testing in England, concentrating on the Counties of Norfolk and Kent, where Ropers from Norfolk migrated.

From CL of Savannah, GA

"I used FamilyTreeDNA for our guys. They do an extra strand, I believe. Our Rosenblaths go back to the 1600's.However when they came to America their name was misspelled many different ways. It was hard to tell who was related to whom. Recently we had a confirmed "Rosenplot" relation in TN, originating in Canada as with my Grandmother who is descended from Rosenplot, submitted his DNA test to see if he was related to an unknown Rosenblath in Shrevesport LA.

"It comes to find out we have several relations, a Robertson who thought they were Scottish because of their name, and several more differently spelled Rosenplots/Rosenblaths. It has brought all of us who were strangers together in a very meaningful way.We share news about our families. Some have been able to 'paper' prove the connection while others are still looking for a few generations old grandfather in the tree we have. It has been great! A lot of new contacts!"

CHAPTER 16

▼

CONSUMER'S GUIDE TO
DNA

"The human genome is about 3 x 109 base pairs long, which would weigh about 40 pg picograms: 1 pg=10-12 grams) per genome," reports Michael Onken, and this description appears on the science Web site of Ricky J. Sethi, MadSci.ORG Administrator at MadSci.ORG

(http://www.madsci.org) at the Washington University School of Medicine, http://www.madsci.org/. "Human cells are diploid, i.e. each contains two copies of the genome, so the nuclear DNA from a human cell would weigh about 80 pg. If we want total cellular DNA, then we need to include mitochondrial DNA (mtDNA).

"The human mitochondrial genome is about 16,000 base pairs long. There are about 10 copies of the genome per mitochondrion, and there are on the order of 1,000 mitochondria per cell. This gives us about 0.2 pg of mtDNA per human cell.

"There are on the order of 1014 cells per adult human, many of which are without nuclei, like skin cells and red blood cells. This would give us just under a kilogram of chromosomal DNA and on the order of a few grams of mitochondrial DNA in the average human body."

(This excerpt is reprinted with permission of Ricky J. Sethi, MadSci.ORG Administrator, at the Washington University School of Medicine. See the Web site at http://www.madsci.org/.)

Knowing how many genes a human has in the future will help not only genealogists and other family and oral historians trace ancestors and keep records of lineages, but physicians will be able to tailor medicines to help people based on how their individual genes react to different elixirs, drugs, natural supplements, herbs, foods, and medicines.

Combined with the knowledge of rainforest tropical plants and their cures, the human genome is headed towards individualization and customization, with an appropriate mixture of food, medicine, or therapy based on one's individual genetic makeup.

To the person without a science background, knowing one's genes also is a way to connect people to their common ancestors in the past and to those descendants. Family history can be researched not only for medical reasons, but for historical reasons, and to show how people are related to one another down through the ages.

To understand how DNA testing relates to history and family records, let's look at some basics of genomics such as what are cells and what is DNA. Then we can think about ways we can use the results of DNA tests in the realm of family history.

Credit for the following Primer below and Dictionary at the back of this book is acknowledged to the U.S. Department of Energy Human Genome Program as the source for both and included also here is the U.S. Department of Energy Human Genome Program's Web site for more information on the Human Genome Project and its applications: www.ornl.gov/hgmis.

This document may be cited in the following style: Human Genome Program, U.S. Department of Energy, Genomics and Its Impact on Medicine and Society: A 2001 Primer, 2001. For printed copies, please contact Laura Yust at Oak Ridge National Laboratory. Send questions or comments to the author, Denise K. Casey. Site on the Web designed by Marissa Mills. This primer was prepared by Denise Casey, Human Genome Management Information System, Oak Ridge National Laboratory. You can find this Primer on the Web at: http://www.ornl.gov/hgmis/publicat/primer2001/1.html and the index to additional publications at:
http://www.ornl.gov/hgmis/publicat/primer2001/index.html

Genomics and Its Impact on Medicine and Society: A 2001 Primer (Courtesy of the U.S. Department of Energy Human Genome Program: http://www.ornl.gov/hgmis

The Basics

Cells are the fundamental working units of every living system. All the instructions needed to direct their activities are contained within the chemical DNA (deoxyribonucleic acid).

DNA from all organisms is made up of the same chemical and physical components. The DNA sequence is the particular side-by-side arrangement of bases along the DNA strand (e.g., ATTCCGGA). This order spells out the exact instructions required to create a particular organism with its own unique traits.

The genome is an organism's complete set of DNA. Genomes vary widely in size: the smallest known genome for a free-living organism (a bacterium) contains about 600,000 DNA base pairs, while human and mouse genomes have some 3 billion. Except for mature red blood cells, all human cells contain a complete genome.

DNA in the human genome is arranged into 24 distinct chromosomes—physically separate molecules that range in length from about 50 million to 250 million base pairs. A few types of major chromosomal abnormalities, including missing or extra copies or gross breaks and rejoinings (translocations), can be detected by microscopic examination. Most changes in DNA, however, are more subtle and require a closer analysis of the DNA molecule to find perhaps single-base differences.

Each chromosome contains many genes, the basic physical and functional units of heredity. Genes are specific sequences of bases that encode instructions on how to make proteins. Genes comprise only about 2% of the human genome; the remainder consists of noncoding regions, whose functions may include providing chromosomal structural integrity and regulating where, when, and in what quantity proteins are made. The human genome is estimated to contain 30,000 to 40,000 genes, but could contain around 70,000 genes. The exact figure is not fully agreed upon yet.

Although genes get a lot of attention, it's the proteins that perform most life functions and even make up the majority of cellular structures. Proteins are large, complex molecules made up of smaller subunits called amino acids.

Chemical properties that distinguish the 20 different amino acids cause the protein chains to fold up into specific three-dimensional structures that define their particular functions in the cell.

The constellation of all proteins in a cell is called its proteome. Unlike the relatively unchanging genome, the dynamic proteome changes from minute to minute in response to tens of thousands of intra-and extra cellular environmental signals.

A protein's chemistry and behavior are specified by the gene sequence and by the number and identities of other proteins made in the same cell at the same time and with which it associates and reacts. Studies to explore protein structure and activities, known as proteomics, will be the focus of much research for decades to come and will help elucidate the molecular basis of health and disease.

CHAPTER 17

▼

HUMAN GENOMIC
PROJECT

Reprinted with permission of the US Dept. of Energy, Human Genome Program, http://www.ornl.gov.hgmis

Human Genome Project

Genomics and Its Impact on Medicine and Society: A 2001 Primer

A Little Bit of History

Though surprising to many, the Human Genome Project (HGP) traces its roots to an initiative in the U.S. Department of Energy (DOE). Since 1945, DOE and its predecessor agencies have been charged by Congress to develop new energy resources and technologies and to pursue a deeper understanding of potential health and environmental risks posed by their production and use. Such studies have since provided the scientific basis for individual risk assessments of nuclear medicine technologies, for example.

In 1986, DOE took a bold step in announcing its Human Genome Initiative, convinced that DOE's missions would be well served by a reference human genome sequence. Shortly there-after, DOE and the National Institutes of Health developed a plan for a joint HGP that officially began in 1990.

Ambitious Goals...

From the outset, the HGP's ultimate goal has been to generate a high-quality reference sequence for the entire human genome and to identify all human genes. Other important goals are to sequence the genomes of model organisms to help interpret human DNA, enhance computational resources to support future research and commercial applications, and explore gene function through mouse-human comparisons. Potential applications are numerous and include customized medicines, improved agricultural products, new energy resources, and tools for environmental cleanup. The HGP also aims to train future scientists, study human variation, and address critical societal issues arising from the increased availability of personal human genome data and related analytical technologies.

...and Exciting Progress

Although the HGP originally was planned to last 15 years, rapid technological advances and worldwide participation have accelerated the expected completion date to 2003. In June 2000, scientists announced biology's most stunning achievement: the generation of a working draft sequence of the entire human genome. In addition to serving as a scaffold for the finished version, the draft provides a road map to an estimated 90% of genes on every chromosome and already has enabled gene hunters to pinpoint genes associated with more than 30 disorders.

HGP Spinoff Projects

- **Microbial Genome Project**
  - www.sc.doe.gov/ober/microbial.html
  - www.ornl.gov/microbialgenomes/
- **Microbial Cell Project**
  microbialcellproject.org
- **Genomes to Life**
  doegenomestolife.org

HGP resources have spurred a boom in spin-off sequencing programs on the human and other genomes in both the private and public sectors. To stimulate further research, all data generated in the public sector are made available rapidly and free of charge via the Web.

- **Environmental Genome Project**
  www.niehs.nih.gov/envgenom/
  home.htm

- **Cancer Genome Anatomy Project**
  www.ncbi.nlm.nih.gov/ncicgap/

- **SNP Consortium**
  snp.cshl.org

Human Genome Project Goals 1998-2003

Human DNA Sequencing

The HGP's continued emphasis is on obtaining by 2003 a complete and highly accurate reference sequence (1 error in 10,000 bases) that is largely continuous across all human chromosomes. Scientists believe that knowing this sequence is critically important for understanding human biology and for applications to other fields.

A "working draft" of the sequence was completed 18 months ahead of schedule, in June 2000. The achievement has provided scientists worldwide with a road map to an estimated 90% of genes on every chromosome. Although the draft contains gaps and errors and does not yet meet the standard of accuracy outlined above, it provides a valuable scaffold for generating a high-quality reference genome sequence. HGP scientists make human DNA sequence available broadly, rapidly, and free of charge via the Web.

Sequencing Technology

Although current sequencing capacity is far greater than at the inception of the HGP, further incremental progress in sequencing technologies, efficiency, and cost-reduction are needed. For future sequencing applications, planners emphasize the importance of supporting novel technologies that may be 5 to 10 years in development.

Sequence Variation

Although more than 99% of human DNA sequences are the same across the population, variations in DNA sequence can have a major impact on how humans respond to disease; to such environmental insults as bacteria, viruses, toxins, and chemicals; and to drugs and other therapies.

Methods are being developed to detect different types of variation, particularly the most common type called single-nucleotide polymorphisms (SNPs), which occur about once every 100 to 300 bases. Scientists believe SNP maps will help them identify the multiple genes associated with such complex diseases as cancer, diabetes, vascular disease, and some forms of mental illness. These associations are difficult to establish with conventional gene-hunting methods because a single altered gene may make only a small contribution to disease risk.

Functional Genomics

Efficient interpretation of the functions of human genes and other DNA sequences requires that strategies be developed to enable large-scale investigations across whole genomes. A first priority is to generate complete sets of full-length cDNA clones and sequences for human and model-organism genes. Other functional-genomics goals include studies into gene expression and control and the development of experimental and computational methods for understanding gene function.

Comparative Genomics

The functions of human genes and other DNA regions often are revealed by studying their parallels in nonhumans. HGP researchers have obtained complete genomic sequences for the bacterium Escherichia coli, the yeast Saccharomyces cerevisiae, the fruit fly Drosophila melanogaster, and the roundworm Caenorhabditis elegans. Sequencing continues on the laboratory mouse. The availability of complete genome sequences generated both inside and outside the HGP is driving a major breakthrough in fundamental biology as scientists compare entire genomes to gain new insights into evolutionary, biochemical, genetic, metabolic, and physiological pathways.

Ethical, Legal, and Social Implications (ELSI)
Rapid advances in the science of genetics and its applications present new and complex ethical and policy issues for individuals and society. ELSI programs that identify and address these implications have been an integral part of the U.S. HGP since its inception. These programs have resulted in a body of work that promotes education and helps guide the conduct of genetic research and the development of related medical and public policies.

Bioinformatics and Computational Biology
Continued investment in current and new databases and analytical tools is critical to the success of the HGP and to the future usefulness of the data it produces. Databases must adapt to the evolving needs of the scientific community and must allow queries to be answered easily. Planners suggest developing a human genome database, analogous to model organism databases that will link to phenotypic information. Also needed are databases and analytical tools for studying the expanding body of gene-expression and functional data, for modeling complex biological networks and interactions, and for collecting and analyzing sequence-variation data.

Training
Future genome scientists will require training in interdisciplinary areas including biology, computer science, engineering, mathematics, physics, and chemistry. Also, scientists with management skills will be needed for leading large data-production efforts.

See Previous Goals
1998-2003 Third Five Year Plan
1993-1998 Second Five Year Plan
1991-1995 Original HGP Goals

This document may be cited in the following style:
Human Genome Program, U.S. Department of Energy, *Genomics and Its Impact on Medicine and Society: A 2001 Primer*, 2001.

For printed copies, please contact <u>Laura Yust</u> at Oak Ridge National Laboratory. Send questions or comments to the author, <u>Denise K. Casey</u>. Site designed by Marissa Mills.

CHAPTER 18

▼

WHAT WE'VE LEARNED SO FAR

Genomics and Its Impact on Medicine and Society: A 2001 Primer

Achievement of a Draft Human Genome Sequence

Reprinted with permission of the US Dept. of Energy, Human Genome Program, http://www.ornl.gov.hgmis

In February 2001, HGP and Celera Genomics scientists published the long-awaited details of the working-draft DNA sequence. Although the draft is filled with mysteries, the first panoramic view of the human genetic landscape has revealed a wealth of information and some early surprises. Papers describing research observations in the journals *Nature* (Feb. 15, 2001) and *Science* (Feb. 16, 2001) are freely accessible.

Although clearly not a Holy Grail or Rosetta Stone for deciphering all of biology—two early metaphors commonly used to describe the coveted prize—the sequence is a magnificent and unprecedented resource that will serve as a basis for research and discovery throughout this century and beyond. It will have diverse practical applications and a profound impact upon how we view ourselves and our place in the tapestry of life.

One insight already gleaned from the sequence is that, even on the molecular level, we are more than the sum of our 35,000 or so genes. Surprisingly, this newly estimated number of genes is only one-third as great as previously thought and only twice as many as those of a tiny transparent worm, although the numbers may be revised as more computational and experimental analyses are performed.

At once humbled and intrigued by this finding, scientists suggest that the genetic key to human complexity lies not in the number of genes but in how gene parts are used to build different products in a process called alternative splicing. Other sources of added complexity are the thousands of chemical modifications made to proteins and the repertoire of regulatory mechanisms controlling these processes.

The draft encompasses 90% of the human genome's euchromatic portion, which contains the most genes. In constructing the working draft, the 16 genome sequencing centers produced over 22.1 billion bases of raw sequence data, comprising overlapping fragments totaling 3.9 billion bases and providing sevenfold coverage (sequenced seven times) of the human genome. Over 30% is high-quality, finished sequence, with eight-to tenfold coverage, 99.99% accuracy, and few gaps.

High-Quality Version Expected in 2003

The entire working draft will be finished to high quality by 2003. Coincidentally, that year also will be the 50th anniversary of Watson and Crick's <u>publication of DNA structure</u> that launched the era of molecular genetics. Much will remain to be deciphered even then. Some highlights follow from Nature, Science, and The Wellcome Trust philanthropy (an HGP funder).

By the Numbers

- The human genome contains 3164.7 million chemical nucleotide bases (A, C, T, and G).

- The average gene consists of 3000 bases, but sizes vary greatly, with the largest known human gene being dystrophin at 2.4 million bases.

- The total number of genes is estimated at 30,000 to 40,000, much lower than previous estimates of 80,000 to 140,000 that had been based on extrapolations from gene-rich areas as opposed to a composite of gene-rich and gene-poor areas.

- The order of almost all (99.9%) nucleotide bases is exactly the same in all people.

- The functions are unknown for more than 50% of discovered genes.

The Wheat from the Chaff

- About 2% of the genome encodes instructions for the synthesis of proteins.

- Repeated sequences that do not code for proteins (Òjunk DNAÓ) make up at least 50% of the human genome.

- Repetitive sequences are thought to have no direct functions, but they shed light on chromosome structure and dynamics. Over time, these repeats reshape the genome by rearranging it, thereby creating entirely new genes or modifying and reshuffling existing genes.

- During the past 50 million years, a dramatic decrease seems to have occurred in the rate of accumulation of these repeats.

How It's Arranged

- The human genome's gene-dense "urban centers" are composed predominantly of the DNA building blocks G and C.

- In contrast, the gene-poor "deserts" are rich in the DNA building blocks A and T. GC-and AT-rich regions usually can be seen through a microscope as light and dark bands on the chromosomes.

- Genes appear to be concentrated in random areas along the genome, with vast expanses of noncoding DNA between.

- Stretches of up to 30,000 C and G bases repeating over and over often occur adjacent to gene-rich areas, forming a barrier between the genes and the "junk DNA." These CpG islands are believed to help regulate gene activity.

- Chromosome 1 has the most genes (2968), and the Y chromosome has the fewest (231).

How the Human Genome Compares with Those of Other Organisms

- Unlike the human's seemingly random distribution of gene-rich areas, many other organisms' genomes are more uniform, with genes evenly spaced throughout.

- Humans have on average three times as many kinds of proteins as the fly or worm because of mRNA transcript "alternative splicing" and chemical modifications to the proteins. This process can yield different protein products from the same gene.

- Humans share most of the same protein families with worms, flies, and plants, but the number of gene family members has expanded in humans, especially in proteins involved in development and immunity.

- The human genome has a much greater portion (50%) of repeat sequences than the mustard weed (11%), the worm (7%), and the fly (3%).

- Although humans appear to have stopped accumulating repetitive DNA over 50 million years ago, there seems to be no such decline in rodents. This may account for some of the fundamental differences between hominids and rodents, although estimates of gene numbers are similar in both species. Scientists have proposed many theories to explain evolutionary contrasts between humans and other organisms, including life span, litter sizes, inbreeding, and genetic drift.

Variations and Mutations

- Scientists have identified about 1.4 million locations where single-base DNA differences (SNPs, see Goals Box: Sequence Variation) occur in humans. This information promises to revolutionize the processes of finding chromosomal locations for disease-associated sequences and tracing human history.

- The ratio of germline (sperm or egg cell) mutations is 2:1 in males vs females. Researchers point to several reasons for the higher mutation rate in the male germline, including the greater number of cell divisions required for sperm formation than for eggs.

Applications, Future Challenges

Deriving meaningful knowledge from the DNA sequence will define research through the coming decades to inform our understanding of biological systems. This enormous task will require the expertise and creativity of tens of thousands of scientists from varied disciplines in both the public and private sectors worldwide.

The draft sequence already is having an impact on finding genes associated with disease. Genes have been pinpointed and associated with numerous diseases and disorders including breast cancer, muscle disease, deafness, and blindness. Additionally, finding the DNA sequences underlying such common diseases as cardiovascular disease, diabetes, arthritis, and cancers is being aided by the human SNP maps generated in the HGP in cooperation with the private sector. These genes and SNPs provide focused targets for the development of effective new therapies.

One of the greatest impacts of having the sequence may well be in enabling an entirely new approach to biological research. In the past, researchers studied one or a few genes at a time. With whole-genome sequences and new automated, high-throughput technologies, they can approach questions systematically and on a grand scale. They can study all the genes in a genome, for example, or all the gene products in a particular tissue or organ or tumor, or how tens of thousands of genes and proteins work together in interconnected networks to orchestrate the chemistry of life.

CHAPTER 19

▼

AFTER THE HUMAN GENOME PROJECT (HGP) THE NEXT STEPS...

Genomics and Its Impact on Medicine and Society: A 2001 Primer

The words of Winston Churchill, spoken in 1942 after 3 years of war, capture well the HGP era: "Now this is not the end. It is not even the beginning of the end. But it is, perhaps, the end of the beginning."

The avalanche of genome data grows daily. The new challenge will be to use this vast reservoir of data to explore how DNA and proteins work with each other and the environment to create complex, dynamic living systems.

Systematic studies of function on a grand scale—functional genomics—will be the focus of biological explorations in this century and beyond. These explorations will encompass studies in transcriptosmics, proteomics, structural genomics, new experimental methodologies, and comparative genomics.

- **Transcriptomics** involves large-scale analysis of messenger RNAs (molecules that are transcribed from active genes) to determine when, where, and under what conditions genes are expressed.

- **Proteomics**—the study of protein expression and function—can bring researchers closer than gene-expression studies to what's actually happening in the cell.

- **Structural genomics** initiatives are being launched worldwide to generate the 3-D structures of one or more proteins from each protein family, thus offering clues to their function and providing biological targets for drug design.

- **Knockout studies** are one experimental method for understanding the function of DNA sequences and the proteins they encode. Researchers inactivate genes in living organisms and monitor any changes that could reveal the function of specific genes.

- **Comparative genomics**—analyzing DNA sequence patterns of humans and well-studied model organisms side by side—has become one of the most powerful strategies for identifying human genes and interpreting their function.

Medicine and the New Genetics: Gene Testing, Pharmacogenomics, and Gene Therapy

DNA underlies every aspect of human health, both in function and dysfunction. Obtaining a detailed picture of how genes and other DNA sequences function together and interact with environmental factors ultimately will lead to the discovery of pathways involved in normal processes and in disease pathogenesis. Such knowledge will have a profound impact on the way disorders are diagnosed, treated, and prevented and will bring about revolutionary changes in clinical and public health practice. Some of these transformative developments are described below.

Gene Tests

DNA-based tests are among the first commercial medical applications of the new genetic discoveries. Gene tests can be used to diagnose disease, confirm a diagnosis, provide prognostic information about the course of disease, confirm the existence of a disease in asymptomatic individuals, and, with varying degrees of accuracy, predict the risk of future disease in healthy individuals or their progeny.

Currently, several hundred genetic tests are in clinical use, with many more under development, and their numbers and varieties are expected to increase rapidly over the next decade. Most current tests detect mutations associated with rare genetic disorders that follow Mendelian inheritance patterns. These include myotonic and Duchenne muscular dystrophies, cystic fibrosis, neurofibromatosis type 1, sickle cell anemia, and Huntington's disease.

Recently, tests have been developed to detect mutations for a handful of more complex conditions such as breast, ovarian, and colon cancers. Although they have limitations, these tests sometimes are used to make risk estimates in pre-symptomatic individuals with a family history of the disorder.

One potential benefit to using these gene tests is that they could provide information that helps physicians and patients manage the disease or condition more effectively. Regular colonoscopies for those having mutations associated with colon cancer, for instance, could prevent thousands of deaths each year.

Some scientific limitations are that the tests may not detect every mutation associated with a particular condition (many are as yet undiscovered), and the ones they do detect may present different risks to different people and populations. Another important consideration in gene testing is the lack of effective treatments or preventive measures for many diseases and conditions now being diagnosed or predicted.

Revealing information about risk of future disease can have significant emotional and psychological effects as well. Moreover, the absence of privacy and legal protections can lead to discrimination in employment or insurance or other misuse of personal genetic information. Additionally, because genetic tests reveal information about individuals and their families, test results can affect family dynamics. Results also can pose risks for population groups if they lead to group stigmatization.

Other issues related to gene tests include their effective introduction into clinical practice, the regulation of laboratory quality assurance, the availability of testing for rare diseases, and the education of healthcare providers and patients about correct interpretation and attendant risks.

Pharmacogenomics: Moving Away from "One-Size-Fits-All" Therapeutics

Within the next decade, researchers will begin to correlate DNA variants with individual responses to medical treatments, identify particular subgroups of patients, and develop drugs customized for those populations. The discipline that blends pharmacology with genomic capabilities is called pharmacogenomics.

More than 100,000 people die each year from adverse responses to medications that are beneficial to others. Another 2.2 million experience serious reactions, while others fail to respond at all. DNA variants in genes involved in drug metabolism, particularly the cytochrome P450 multigene family, are the focus of much current research in this area.

Enzymes encoded by these genes are responsible for metabolizing most drugs used today, including many for treating psychiatric, neurological, and cardiovascular diseases. Enzyme function affects patients' responses to both the drug and the dose. Future advances will enable rapid testing to determine the patient's genotype and drastically reduce hospitalization resulting from adverse reactions.

Genomic data and technologies also are expected to make drug development faster, cheaper, and more effective. Most drugs today are based on about 500 molecular targets; genomic knowledge of the genes involved in diseases, disease pathways, and drug-response sites will lead to the discovery of thousands of new targets.

New drugs, aimed at specific sites in the body and at particular biochemical events leading to disease, probably will cause fewer side effects than many current medicines. Ideally, the new genomic drugs could be given earlier in the disease process. As knowledge becomes available to select patients most likely to benefit from a potential drug, pharmacogenomics will speed the design of clinical trials to bring the drugs to market sooner.

Gene Therapy, Enhancement

The potential for using genes themselves to treat disease or enhance particular traits has captured the imagination of the public and the biomedical community. This largely experimental field—gene transfer or gene therapy—holds potential for treating or even curing such genetic and acquired diseases as cancers and AIDS by using normal genes to supplement or replace defective genes or bolster a normal function such as immunity.

More than 500 clinical gene-therapy trials involving about 3500 patients have been identified worldwide (June 2001). The vast majority (78%) take place in the United States, followed by Europe (18%). Although most trials focus on various types of cancer, studies also involve other multigenic and monogenic, infectious, and vascular diseases. Protocols generally are aimed at establishing the safety of gene-delivery procedures rather than effectiveness, and no cures as yet can be attributed to these trials.

Gene transfer still faces many scientific obstacles before it can become a practical approach for treating disease. According to the American Society of Human Genetics' Statement on Gene Therapy, effective progress will be achieved only through continued rigorous research on the most fundamental mechanisms underlying gene delivery and gene expression in animals.

CHAPTER 20

▼

SOCIETAL CONCERNS ARISING FROM THE NEW GENETICS

Genomics and Its Impact on Medicine and Society: A 2001 Primer

Since its inception, the Human Genome Project has dedicated funds toward studying the ethical, legal, and social issues surrounding the availability of the new data and capabilities. Examples of such issues follow.

- **Privacy and confidentiality of genetic information.** *Who owns and controls genetic information? Is genetic privacy different from medical privacy?*

- **Fairness in the use of genetic information by insurers, employers, courts, schools, adoption agencies, and the military, among others.** *Who should have access to personal genetic information, and how will it be used?*

- **Psychological impact, stigmatization, and discrimination due to an individual's genetic differences.** *How does personal genetic information affect self-identity and society's perceptions?*

- **Reproductive issues including adequate and informed consent and the use of genetic information in reproductive decision making.** *Do healthcare personnel properly counsel parents about risks and limitations? What are the larger societal issues raised by new reproductive technologies?*

- **Clinical issues including the education of doctors and other health-service providers, people identified with genetic conditions, and the general public; and the implementation of standards and quality-control measures.** *How should health professionals be prepared for the new genetics? How can the public be educated to make informed choices? How will genetic tests be evaluated and regulated for accuracy, reliability, and usefulness?* (Currently, there is little regulation at the federal level.) *How does society balance current scientific limitations and social risk with long-term benefits?*

- **Fairness in access to advanced genomic technologies.** *Who will benefit? Will there be major worldwide inequities?*

- **Uncertainties associated with gene tests for susceptibilities and complex conditions (e.g., heart disease, diabetes, and Alzheimer's disease).** *Should testing be performed when no treatment is available or when interpretation is unsure? Should children be tested for susceptibility to adult-onset diseases?*

- **Conceptual and philosophical implications regarding human responsibility, free will vs. genetic determinism, and concepts of health and disease.** *Do our genes influence our behavior, and can we control it? What is considered acceptable diversity? Where is the line drawn between medical treatment and enhancement?*

- **Health and environmental issues concerning genetically modified (GM) foods and microbes.** *Are GM foods and other products safe for humans and the environment? How will these technologies affect developing nations' dependence on industrialized nations?*

Commercialization of products including property rights (patents, copyrights, and trade secrets) and accessibility of data and materials. *Will patenting DNA sequences limit their accessibility and development into useful products?*

CHAPTER 21

▼

How to Interview Older Adults for Intergenerational Life Story Writing and Memory Enhancement: Putting the Events into the Time Capsule

Working with Intergenerational Writing or Older Writers

You've put the DNA test results with interpretations, the prescribed foods and nutraceuticals in the time capsule. You've included the ancestry DNA tests. Next, you need the memories in text, pictures, sound, and memories. Whether narration, story, or skit, you need the turning points and the highlights of the events.

STEP 1: Visit places yourself where seniors live and congregate socially or send students of various ages from high school to graduate school in small groups to visit senior community centers, nursing homes, adult education classrooms in lifestory writing, extended studies programs at universities for older adults in lifelong learning, religious services social halls, or senior apartment complexes activity rooms.

STEP 2: Bring tape recorders and camcorders and carry note pads to write down important points. You can also use visuals such as photo scrapbooks for inspiring and refreshing memoires as older adults talk about the turning points and highlights of their life stories. As they recall events and their part in those evens, their memories are working. They are visualizing.

STEP 3: Assign each student one or two older persons to interview with the following questions.

1. What were the most significant turning points or events in your life?

2. How did you survive the Wars?

3. What were the highlights, turning points, or significant events that you experienced during the economic downturn of 1929-1939? How did you cope or solve your problems?

4. What did you do to solve your problems during the significant stages of your life at age 10, 20, 30, 40, 50, 60 and 70-plus? Or pick a year that you want to talk about.

5. What changes in your life do you want to remember and pass on to future generations?

6. What was the highlight of your life?

7. How is it best to live your life after 70?

8. What years do you remember most?

9. What was your favorite stage of life?

10. What would you like people to remember about you and the times you lived through?

STEP 3: Have the student record the older person's answers. Select the most significant events, experiences, or turning points the person chooses to emphasize. Then write the story of that significant event in ten pages or less.

STEP 4: Ask the older person to supply the younger student photos, art work, audio tapes, or video clips. Usually photos, pressed flowers, or art work will be supplied. Have the student or teacher scan the photos onto a disk and return the original photos or art work or music to the owner.

STEP 5:The student and/or teacher scans the photos and puts them onto a Web site on the Internet at one of the free communities that give away Web site to the public at no cost...some include http://www.tripod.com, http://www.fortunecity.com, http://www.angelfire.com, http://www.geocities.com, and others. Most search engines will give a list of communities at offering free Web sites to the public. Microsoft also offers free family Web sites for family photos and newsletters or information. Ask your Internet service provider whether it offers free Web site space to subscribers.

1. Create a Web site with text from the older person's significant life events

2. Add photos.

3. Add sound or.wav files with the voice of the older person speaking in small clips or sound bites.

4. Intersperse text and photos or art work with sound, if available.
    Add video clips, if available and won't take too much bandwidth.

5. Put Web site on line as TIME CAPSULE of (insert name of person) interviewed and edited by, insert name of student who interviewed older person.

STEP 6: Label each Web site Time Capsule and collect them in a history archives on the lives of older adults at the turn of the millennium. Make sure the older person and all relatives and friends are emailed the Web site link. You have now created a time capsule for future generations.

This can be used as a classroom exercise in elementary and high schools to teach the following:

1. Making friends with older adults.

2. Learning to write on intergenerational topics.

3. Bringing community together of all generations.

4. Learning about foster grandparents.

5. History lessons from those who lived through history.

6. Learning about diversity and how people of diverse origins lived through the 20th century.

7. Preserving the significant events in the lives of people as time capsules for future generations to know what it was like to live between 1900 and 2000 at any age.

8. Learning to write skits and plays from the life stories of older adults taken down by young students.

9. Teaching older adults skills in creative writing at senior centers.

10. Learning what grandma did during World War 2 or the stock market crash of 1929 followed by the economic downturn of 1930-1938.

What to Ask People about Their Lives

Step 1
When you interview, ask for facts and concrete details. Look for statistics, and research whether statistics are deceptive in your case.

Step 2
To write a plan, write one sentence for each topic that moves the story or piece forward. Then summarize for each topic in a paragraph. Use dialogue at least in every third paragraph.

Step 3
Look for the following facts or headings to organize your plan for a biography or life story.

1.   SLOGAN. Ask the people you interview what would be their slogan if they had to create/invent a slogan that fit themselves or their aspirations: One slogan might be something like the seventies ad for cigarettes, "We've come a long way, baby," to signify ambition. Only look for an original slogan.

2.   CRUSADE. Ask the people you interview or a biography, for what purpose is or was their crusade? Is or was it equality in the workplace or something personal and different such as dealing with change—downsizing, working after retirement, or anything else?

3.   IMPACT. Ask what makes an impact on people's lives and what impact the people you're interviewing want to make on others?

4.   STATISTICS: How deceptive are they? How can you use them to focus on reality?

5.   How have the people that you're interviewing influenced changes in the way people or corporations function?

6.   To what is the person aspiring?

7.   What kind of communication skills does the person have and how are these skills received? Are the communication skills male or female, thinking or feeling, yin or yang, soft or steeled, and are people around these people negative or positive about those communication skills?

8.   What new styles is the person using? What kind of motivational methods, structure, or leadership? Is the person a follower or leader? How does the person match his or her personality to the character of a corporation or interest?

9.   How does the person handle change?

10. How is the person reinforced?

Once you have titles and summarized paragraphs for each segment of your story, you can more easily flesh out the story by adding dialogue and description to your factual information. Look for differences in style between the people you interview? How does the person want to be remembered?

Is the person a risk taker or cautious for survival? Does the person identify with her job or the people involved in the process of doing the work most creatively or originally? Does creative expression take precedence over processes of getting work out to the right place at the right time? Does the person want his ashes to spell the word "love" where the sea meets the shore?

Search the Records in the Family History Library of Salt Lake City, Utah

Make use of the database online at the Family History Library of Salt Lake City, Utah. Or visit the branches in many locations. The Family History Library (FHL) is known worldwide as the focal point of family history records preservation.

The FHL collection contains more than 2.2 million rolls of microfilmed genealogical records, 742,000 microfiche, 300,000 books, and 4,500 periodicals that represent data collected from over 105 countries. You don't have to be a member of any particular church or faith to use the library or to go online and search the records.

Family history records owe a lot to the invention of writing. And then there is oral history, but someone needs to transcribe oral history to record and archive them for the future.

Interestingly, isn't it a coincidence that writing is 6,000 years old and DNA that existed 6,000 years ago first reached such crowded conditions in the very cities that had first used writing extensively to measure accounting and trade had very little recourse but to move on to new areas where there were far less people and less use of writing?

A lot of major turning points occurred 6,000 years ago—the switch to a grain-based diet from a meat and root diet, the use of bread and fermented grain beverages, making of oil from plants, and the rise of religions based on building "god houses" in the centers of town in areas known as the "cereal belt" around the world.

Six thousand years ago in India we have the start of the Sanksrit writings, the cultivation of grain. In China, we have the recording of acupuncture points for medicine built on energy meridians that also show up in the blue tattoos of the Ice Man fossil "Otsi" in the Alps—along the same meridians as the Chinese acupuncture points.

At 6,000 years ago the Indo European languages spread out across Europe. Mass migrations expanded by the Danube leaving pottery along the trade routes that correspond to the clines and gradients of gene frequency coming out of the cereal belts.

Then something happened. There was an agricultural frontier cutting off the agriculturists from the hunters. Isn't it a coincidence that the agricultural frontiers or barriers also are genetic barriers at least to some degree?

CHAPTER 22

▼

ORAL HISTORY—EVEN EATING CUSTOMS ARE PART OF ORAL HISTORY AND TIME CAPSULES

Here's how to systematically collect, record, and preserve living peoples' testimonies about their own experiences. After you record in audio and/or video the highlights of anyone's experiences, try to verify your findings. See whether you can check any facts in order to find out whether the person being recorded is making up the story or whether it really did happen.

This is going to be difficult unless you have witnesses or other historical records. Once you have verified your findings to the best of your ability, note whether the findings have been verified. Then analyze what you found. Put the oral history recordings in an accurate historical context.

Mark the recordings with the dates and places. Watch where you store your findings so scholars in the future will be able to access the transcript or recording and convert the recording to another, newer technology. For instance, if you have a transcript on paper, have it saved digitally on a disk and somewhere else on tape and perhaps a written transcript on acid-free good paper in case technology moves ahead before the transcript or recording is converted to the new technology.

For example, if you only put your recording on a phonograph record, within a generation or two, there may not be any phonographs around to play the record. The same goes for CDs, DVDs and audio or video tapes.

So make sure you have a readable paper copy to be transcribed or scanned into the new technology as well as the recordings on disk and tape. For example, if you record someone's experiences in a live interview with your video camera, use a cable to save the video in the hard disk of a computer and then burn the file to a CD or DVD. Keep a copy of audio tape and a copy of regular video tape—all in a safe place such as a time capsule, and make a copy for various archives in libraries and university oral history preservation centers. Be sure scholars in the future can find a way to enjoy the experiences in your time capsule, scrapbook, or other storage device for oral histories.

Use your DNA testing results to add more information to a historical record. As an interviewer with a video camera and/or audio tape recorder, your task is to record as a historical record what the person who you are interviewing recollects.

The events move from the person being interviewed to you, the interviewer, and then into various historical records. In this way you can combine results of DNA testing with actual memories of events. If it's possible, also take notes or have someone take notes in case the tape doesn't pick up sounds clearly.

I had the experience of having a video camera battery go out in spite of all precautions when I was interviewing someone, and only the audio worked. So keep a backup battery on hand whether you use a tape recorder or a video camera. If at all possible, have a partner bring a spare camera and newly recharged battery. A fully charged battery left overnight has a good chance of going out when you need it.

What Should Go Into an Oral History?

Emphasize the commitment to family and faith. To create readers' and media attention to an oral history, it should have some redemptive value to a universal audience. That's the most important point. Make your oral history simple and earthy. Write about real people who have values, morals, and a faith in something greater than themselves that is equally valuable to readers or viewers.

Publishers who buy an oral history written as a book on its buzz value are buying simplicity. It is simplicity that sells and nothing else but simplicity. This is true for oral histories, instructional materials, and fiction. It's good storytelling to say it simply.

Simplicity means the oral history or memoirs book or story gives you all the answers you were looking for in your life in exotic places, but found it close by. What's the great proverb that your oral history is telling the world?

Is it to stand on your own two feet and put bread on your own table for your family? That's the moral point, to pull your own weight, and pulling your own weight is a buzz word that sells oral histories and fiction that won't preach, but instead teach and reach through simplicity.

That's the backbone of the oral historian's new media. Buzz means the story is simple to understand. You make the complex easier to grasp. And buzz means you can sell your story or book, script or narrative by focusing on the values of simplicity, morals, faith, and universal values that hold true for everyone. Doing the best to take care of your family sells and is buzz appeal, hot stuff in the publishing market of today and in the oral history archives. This is true, regardless of genre. Publishers go through fads every two years—angel books, managing techniques books, computer home-based business books, novels about ancient historical characters or tribes, science fiction, children's programming, biography, and oral history transcribed into a book or play.

The genres shift emphasis, but values are consistent in the bestselling books. Perhaps your oral history will be simple enough to become a bestselling book or script. In the new media, simplicity is buzz along with values. Oral history, like best-selling novels and true stories is built on simplicity, values, morals, and commitment. Include how one person dealt with about trends. Focus your own oral history about life in the lane of your choice.

Steps to Take in Gathering Oral Histories
Use the following sequence when gathering oral/aural histories:

1. Develop one central issue and divide that issue into a few important questions that highlight or focus on that one central issue.

2. Write out a plan just like a business plan for your oral history project. You may have to use that plan later to ask for a grant for funding, if required. Make a list of all your products that will result from the oral history when it's done.

3. Write out a plan for publicity or public relations and media relations. How are you going to get the message to the public or special audiences?

4. Develop a budget. This is important if you want a grant or to see how much you'll have to spend on creating an oral history project.

5. List the cost of video taping and editing, packaging, publicity, and help with audio or special effects and stock shot photos of required.

6. What kind of equipment will you need? List that and the time slots you give to each part of the project. How much time is available? What are your deadlines?

7. What's your plan for a research? How are you going to approach the people to get the interviews? What questions will you ask?

8. Do the interviews. Arrive prepared with a list of questions. It's okay to ask the people the kind of questions they would like to be asked. Know what dates the interviews will cover in terms of time. Are you covering the economic depression of the thirties? World Wars? Fifties? Sixties? Pick the time parameters.

9. Edit the interviews so you get the highlights of experiences and events, the important parts. Make sure what's important to you also is important to the person you interviewed.

10. Find out what the interviewee wants to emphasize perhaps to highlight events in a life story. Create a video-biography of the highlights of one person's life or an oral history of an event or series of events.

11. Process audio as well as video, and make sure you have written transcripts of anything on audio and/or video in case the technology changes or the tapes go bad.

12. Save the tapes to compact disks, DVDs, a computer hard disk and several other ways to preserve your oral history time capsule. Donate any tapes or CDs to appropriate archives, museums, relatives of the interviewee, and one or more oral history libraries. They are usually found at universities that have an oral history department and library such as UC Berkeley and others.

13. Check the Web for oral history libraries at universities in various states and abroad.

14. Evaluate what you have edited. Make sure the central issue and central questions have been covered in the interview. Find out whether newspapers or magazines want summarized transcripts of the audio and/or video with photos.

15. Contact libraries, archives, university oral history departments and relevant associations and various ethnic genealogy societies that focus on the subject matter of your central topic.

16. Keep organizing what you have until you have long and short versions of your oral history for various archives and publications. Contact magazines and newspapers to see whether editors would assign reporters to do a story on the oral history project.

17. Create a scrapbook with photos and summarized oral histories. Write a synopsis of each oral history on a central topic or issue. Have speakers give public presentations of what you have for each person interviewed and/or for the entire project using highlights of several interviews with the media for publicity. Be sure your project is archived properly and stored in a place devoted to oral history archives and available to researchers and authors.

Aural History Techniques

1. Begin with easy to answer questions that don't require you explore and probe deeply in your first question. Focus on one central issue when asking questions.

2. First research written or visual resources before you begin to seek an oral history of a central issue, experience, or event.

3. Who is your intended audience?

4. What kind of population niche or sample will you target?

5. What means will you select to choose who you will interview? What group of people will be central to your interview?

6. Write down how you'll explain your project. Have a script ready so you don't digress or forget what to say on your feet.

7. Consult oral history professionals if you need more information. Make sure what you write in your script will be clear to understand by your intended audience.

8. Have all the equipment you need ready and keep a list of what you'll use and the cost. Work up your budget.

9. Choose what kind of recording device is best—video, audio, multimedia, photos, and text transcript. Make sure your video is broadcast quality. I use a Sony Digital eight (high eight) camera.

10. Make sure from cable TV stations or news stations that what type of video and audio you choose ahead of time is broadcast quality.

11. Make sure you have an external microphone and also a second microphone as a second person also tapes the interview in case the quality of your camera breaks down. You can also keep a tape recorder going to capture the audio in case your battery dies.

12. Make sure your battery is fully charged right before the interview. Many batteries die down after a day or two of nonuse.

13. Test all equipment before the interview and before you leave your office or home. I've had batteries go down unexpectedly and happy there was another person ready with another video camera waiting and also an audio tape version going.

14. Make sure the equipment works if it's raining, hot, cold, or other weather variations. Test it before the interview. Practice interviewing someone on your equipment several times to get the hang of it before you show up at the interview.

15. Make up your mind how long the interview will go before a break and use tape of that length, so you have one tape for each segment of the interview. Make several copies of your interview questions.

16. Be sure the interviewee has a copy of the questions long before the interview so the person can practice answering the questions and think of what to say or even take notes. Keep checking your list of what you need to do.

17. Let the interviewee make up his own questions if he wants. Perhaps your questions miss the point. Present your questions first. Then let him embellish the questions or change them as he wants to fit the central issue with his own experiences.

18. Call the person two days and then one day before the interview to make sure the individual will be there on time and understands how to travel to the location. Or if you are going to the person's home, make sure you understand how to get there.

19. Allow yourself one extra hour in case of traffic jams.

20. Choose a quiet place. Turn off cell phones and any ringing noises. Make sure you are away from barking dogs, street noise, and other distractions.

21. Before you interview make sure the person knows he or she is going to be video and audio-taped.

22. If you don't want anyone swearing, make that clear it's for public archives and perhaps broadcast to families.

23. Your interview questions should follow the journalist's information-seeking format of asking, who, what, where, where, how, and why. Oral history is a branch of journalistic research.

24. Let the person talk and don't interrupt. You be the listener and think of oral history as aural history from your perspective.

25. Make sure only one person speaks without being interrupted before someone else takes his turn to speak.

26. Understand silent pauses are for thinking of what to say.

27. Ask one question and let the person gather his thoughts.

28. Finish all your research on one question before jumping to the next question. Keep it organized by not jumping back to the first question after the second is done. Stay in a linear format.

29. Follow up what you can about any one question, finish with it, and move on to the next question without circling back. Focus on listening instead of asking rapid fire questions as they would confuse the speaker.

30. Ask questions that allow the speaker to begin to give a story, anecdote, life experience, or opinion along with facts. Don't ask questions that can be answered only be yes or no. This is not a courtroom. Let the speaker elaborate with facts and feelings or thoughts.

31. Late in the interview, start to ask questions that explore and probe for deeper answers.

32. Wrap up with how the person solved the problem, achieved results, reached a conclusion, or developed an attitude, or found the answer. Keep the wrap-up on a light, uplifting note.

33. Don't leave the individual hanging in emotion after any intensity of. Respect the feelings and opinions of the person. He or she may see the situation from a different point of view than someone else. So respect the person's right to feel as he does. Respect his need to recollect his own experiences.

34. Interview for only one hour at a time. If you have only one chance, interview for an hour. Take a few minutes break. Then interview for the second hour. Don't interview more than two hours at any one meeting.

35. Use prompts such as paintings, photos, music, video, diaries, vintage clothing, crafts, antiques, or memorabilia when appropriate. Carry the photos in labeled files or envelopes to show at appropriate times in order to prime the memory of the interviewee. For example, you may show a childhood photo and ask "What was it like in that orphanage where these pictures were taken?" Or travel photos might suggest a trip to America as a child, or whatever the photo suggests. For example, "Do you remember when this ice cream parlor inside the ABC movie house stood at the corner of X and Y Street? Did you go there as a teenager? What was your funniest memory of this movie theater or the ice cream store inside back in the fifties?"

36. As soon as the interview is over, label all the tapes and put the numbers in order.

37. A signed release form is required before you can broadcast anything. So have the interviewee sign a release form before the interview.

38. Make sure the interviewee gets a copy of the tape and a transcript of what he or she said on tape. If the person insists on making corrections, send the paper transcript of the tape for correction to the interviewee. Edit the tape as best you can or have it edited professionally.

39. Make sure you comply with all the corrections the interviewee wants changed. He or she may have given inaccurate facts that need to be corrected on the paper transcript.

40. Have the tape edited with the corrections, even if you have to make a tape at the end of the interviewee putting in the corrections that couldn't be edited out or changed.

41. As a last resort, have the interviewee redo the part of the tape that needs correction and have it edited in the tape at the correct place marked on the tape. Keep the paper transcript accurate and up to date, signed with a release form by the interviewee.

42. Oral historians write a journal of field notes about each interview. Make sure these get saved and archived so they can be read with the transcript.

43. Have the field notes go into a computer where someone can read them along with the transcript of the oral history tape or CD.

44. Thank the interviewee in writing for taking the time to do an interview for broadcast and transcript.

45. Put a label on everything you do from the interview to the field notes. Make a file and sub file folders and have everything stored in a computer, in archived storage, and in paper transcript.

46. Make copies and digital copies of all photos and put into the records in a computer. Return originals to owners.

47. Make sure you keep your fingerprints off the photos by wearing white cotton gloves. Use cardboard when sending the photos back and pack securely. Also photocopy the photos and scan the photos into your computer. Treat photos as antique art history in preservation.

48. Make copies for yourself of all photos, tapes, and transcripts. Use your duplicates, and store the original as the master tape in a place that won't be used often, such as a time capsule or safe, or return to a library or museum where the original belongs.

49. Return all original photos to the owners. An oral history archive library or museum also is suitable for original tapes. Use copies only to work from, copy, or distribute.

50. Index your tapes and transcripts. To use oral history library and museum terminology, recordings and transcripts are given "accession numbers."

51. Phone a librarian in an oral history library of a university for directions on how to assign accession numbers to your tapes and transcripts if the materials are going to be stored at that particular library. Store copies in separate places in case of loss or damage.

52. If you don't know where the materials will be stored, use generic accession numbers to label your tapes and transcripts. Always keep copies available for yourself in case you have to duplicate the tapes to send to an institution, museum, or library, or to a broadcast company.

53. Make synopses available to public broadcasting radio and TV stations.

54. Check your facts.

55. Are you missing anything you want to include?

56. Is there some place you want to send these tapes and transcripts such as an ethnic museum, radio show, or TV satellite station specializing in the topics on the tapes, such as public TV stations? Would it be suitable for a world music station? A documentary station?

57. If you need more interviews, arrange them if possible.

58. Give the interviewee a copy of the finished product with the corrections. Make sure the interviewee signs a release form that he or she is satisfied with the corrections and is releasing the tape to you and your project.

59. Store the tapes and transcripts in a library or museum or at a university or other public place where it will be maintained and preserved for many generations and restored when necessary.

60. You can also send copies to a film repository or film library that takes video tapes, an archive for radio or audio tapes for radio broadcast or cable TV.

61. Copies may be sent to various archives for storage that lasts for many generations. Always ask whether there are facilities for restoring the tape. A museum would most likely have these provisions as would a large library that has an oral history library project or section.

62. Make sure the master copy is well protected and set up for long-term storage in a place where it will be protected and preserved.

63. If the oral history is about events in history, various network news TV stations might be interested. Film stock companies may be interested in copies of old photos.

64. Find out from the subject matter what type of archives, repository, or storage museums and libraries would be interested in receiving copies of the oral history tapes and transcripts.

65. Print media libraries would be interested in the hard paper copy transcripts and photos as would various ethnic associations and historical preservation societies. Find out whether the materials will go to microfiche, film, or be digitized and put on CDs and DVDs, or on the World Wide Web. If you want to create a time capsule for the Web, you can ask the interviewee whether he or she wants the materials or selected materials to be put online or on CD as multimedia or other. Then you would get a signed release from the interviewee authorizing you to put the materials or excerpts online. Also find out in whose name the materials are copyrighted and whether you have print and electronic rights to the material or do the owners-authors-interviewees—or you, the videographer-producer? Get it all in writing, signed by those who have given you any interviews, even if you have to call your local intellectual property rights attorney.

How Accurate Are Oral/Aural Histories?

Cameras give fragments, points of view, and bits and pieces. Viewers will see what the videographer or photographer intends to be seen. The interviewee will also be trying to put his point of view across and tell the story from his perspective. Will the photographer or videographer be in agreement with the interviewee? Or if you are recording for print transcript, will your point of view agree with the interviewee's perspective and experience if your basic 'premise,' where you two are coming from, are not in agreement? Think this over as you write your list of questions. Do both of you agree on your central issue on which you'll focus for the interview?

How are you going to turn spoken words into text for your paper hard copy transcript? Will you transcribe verbatim, correct the grammar, or quote as you hear the spoken words? Oral historians really need to transcribe the exact spoken word. You can leave out the 'ahs' and 'oms' or loud pauses, as the interviewee thinks what to say next. You don't want to sound like a court reporter, but you do want to have an accurate record transcribed of what was spoken.

You're also not editing for a movie, unless you have permission to turn the oral history into a TV broadcast, where a lot gets cut out of the interview for time constraints. For that, you'd need written permission so words won't be taken out of context and strung together in the editing room to say something different from what the interviewee intended to say.

Someone talking could put in wrong names, forget what they wanted to say, or repeat themselves. They could mumble, ramble, or do almost anything. So you would have to sit down and weed out redundancy when you can or decide on presenting exactly what you've heard as transcript. When someone reads the transcript in text, they won't have what you had in front of you, and they didn't see and hear the live presentation or the videotape. It's possible to misinterpret gestures or how something is spoken, the mood or tone, when reading a text transcript. Examine all your sources. Use an ice-breaker to get someone talking.

If a woman is talking about female-interest issues, she may feel more comfortable talking to another woman. Find out whether the interviewee is more comfortable speaking to someone of his or her own age. Some older persons feel they can relate better to someone close to their own age than someone in high school, but it varies. Sometimes older people can speak more freely to a teenager.

The interviewee must be able to feel comfortable with the interviewer and know he or she will not be judged. Sometimes it helps if the interviewer is the same ethnic group or there is someone present of the same group or if new to the language, a translator is present.

Read some books on oral history field techniques. Read the National Genealogical Society Quarterly (NGSQ). Also look at The American Genealogist (TAG), The Genealogist, and The New England Historical and Genealogical Register (The Register). If you don't know the maiden name of say, your grandmother's mother, and no relative knows either because it wasn't on her death certificate, try to reconstruct the lives of the males who had ever met the woman whose maiden name is unknown.

Maybe she did business with someone before marriage or went to school or court. Someone may have recorded the person's maiden name before her marriage. Try medical records if any were kept. There was no way to find my mother's grandmother's maiden name until I started searching to see whether she had any brothers in this country. She had to have come as a passenger on a ship around 1880 as she bought a farm. Did her husband come with her?

Was the farm in his name? How many brothers did she have in this country with her maiden surname? If the brothers were not in this country, what countries did they come from and what cities did they live in before they bought the farm in Albany? If I could find out what my great grandmother's maiden name was through any brothers living at the time, I could contact their descendants perhaps and see whether any male or female lines are still in this country or where else on the globe.

Perhaps a list of midwives in the village at the time is recorded in a church or training school for midwives. Fix the person in time and place. Find out whom she might have done business with and whether any records of that business exist. What businesses did she patronize? Look for divorce or court records, change of name records, and other legal documents.

Look at local sources. Did anyone save records from bills of sale for weddings, purchases of homes, furniture, debutante parties, infant supplies, or even medical records? Look at nurses' licenses, midwives' registers, employment contracts, and teachers' contracts, alumni associations for various schools, passports, passenger lists, alien registration cards, naturalization records, immigrant aid societies, city directories, and cross-references. Try religious and women's clubs, lineage and village societies, girl scouts and similar groups, orphanages, sanatoriums, hospitals, police records. Years ago there was even a Eugenics Record Office. What about the women's prisons? The first one opened in 1839—Mount Pleasant Female Prison, NY.

Try voters' lists. If your relative is from another country, try records in those villages or cities abroad. Who kept the person's diaries? Have you checked the Orphan Train records? Try ethnic and religious societies and genealogy associations for that country. Most ethnic genealogy societies have a special interest group for even the smallest villages in various countries.

You can start one and put up a Web site for people who also come from there in past centuries. Check alimony, divorce, and court records, widow's pensions of veterans, adoptions, orphanages, foster homes, medical records, birth, marriage, and death certificates, social security, immigration, pet license owners' files, prisons, alumni groups from schools, passenger lists, military, and other legal records.

When all historical records are being tied together, you can add the DNA testing to link all those cousins. Check military pensions on microfilms in the National Archives. See the bibliography section of this book for further resources on highly recommended books and articles on oral history field techniques and similar historical subjects.

*   *   *   *

Does Writing Your Life Story As A Genealogy and/or Novel Affect Your Memory?

Oral history depends on memory and the ability to speak. I also think of oral history as aural history, based on the ability to hear someone's experiences and remember them to pass on to the next generation or the world.

To find out the effects of oral history on memory and on creative writing on memory, we'd have to ask the people who write their life story and/or genealogy in their older years what it did for them, their memory, and their ability to think and feel. Make use of introverted feeling in writing a commercial or salable life story for the new media. Think in three dimensions for older adults is a different highway. How did DNA testing influence a genealogy search for family history facts?

Did the individual create a time capsule? How was the time capsule saved—on the Web? On a CD, DVD, video, or audio tape? In a scrapbook of photos, with various memorabilia? Did anyone rescue old photos from antique stores and flea markets by searching for photographer's prints on the front or back of the photo or names on the back of the photo and dates or locations?

1. When turning your salable life story, corporate history, or biography into an adventure action romance novel, don't set up your main characters in the first chapter to be in transit traveling on board a plane, train, or ship going somewhere. The action actually starts or hits them after they have already arrived at their destination.

Start your first chapter when your characters already get to their destination place or point in time. A first chapter that opens when your main character is on a plane or train is the kiss of death from many editors point of view and the main reason why a good novel often is rejected. So cut out the traveling scene from your first chapter and begin where the action starts for real, at the destination point. Does anyone visit antique stores, malls, or flea markets to search for family history memorabilia? What about attic, basement, or garage sales?

2. Use a lot of dialogue when turning a biography or your life story into a salable novel, especially in a romance, adventure action, or suspense novel or in one where you combine romance with adventure and suspense.

Use no more than three pages of narrative without dialogue. Let characters speak through the dialogue and tell the reader what is happening. Get characters to speak as normally as possible. If the times and place dictate they speak in proverbs, so be it. Proverbs make the best novels as you turn your proverb into a story and play it out as a novel. Otherwise, have normal speech so you can be the catalyst and bring people together who understand clearly what one another means.

3. Put your characters on the stage and have them talking to one another. If you have introspection in your book, don't use introspection for your action line. Action adventure books keep characters on stage talking to the audience.

4. Use magazines and clothing catalogues to make a collage of what your character might look like. This inspiration may go up on a board in front of you or on the wall to see as you work. Get a picture in your mind of what your characters look like. If they don't exist in art history, draw them yourself or make a mixed media collage of what they look like, speak like, and stand for. Some ideas include the models in "cigar" magazines, catalogues, and fashion publications as well as multi-ethnic and historical illustrations and photos.

5. Research history and keep a loose-leaf notebook with tabs on the history of places you want to research. The history itself is great for ideas on what plot to

write. Look at or visit old forts and similar places. Plug in characters to your research. Look at forts of foreign settlements in the country of your choice, U.S. or any other place. Record the dates in your files. Create a spreadsheet in Excel or any other type of spread sheet with your dates from historical research as these will relate to your characters and help you develop a real plot.

6. Keep a notebook for each novel or biography you write. Put everything related to each book in a notebook. Have one notebook for historical research and one for the novel you're writing or true storybook.

7. When sending out your book manuscript make a media kit for yourself with your resume, photo, list of works in development if you are not yet published, and any other material about your own experience in any other field. Your own biography and photo presented to the press also can be used to let an editor know when you send out your manuscript of what's in development and what you've done.

8. Write down the point of view before your book is begun. Whose point of view is it anyway? Who tells the story? If you're writing a romance novel from your life story or a military romantic suspense novel, true story, radio script, or other genre, agree on the point of view before you start.

9. Who's telling the story and how does she or he know how the other characters know what to say?

10. It's not necessary to continue ethnic stereotypes in your book. If one of your characters is a music agent, for example, and a lot of music agents are of one ethnicity or speak with a certain accent, it's not necessary to continue the stereotyping roles. Pick something new for a change. Otherwise it becomes cliché. Research diverse ways of telling the same story.

• The readers needs to learn facts or experiences, anecdotes, oral histories transcribed, and stories that have not been generalized. Use a series of incidents, action and relationship tension to balance your plot with your dialogue.

• If you're turning a biography into a romance novel, you need to balance the relationship tension with the mystery, action, or other plot. You must have some event occur on both sides, on the sexual tension side and on the mystery or action side to balance out the book.

For every action in a life story, there's an equal and opposite reaction that's primarily character-driven and secondarily plot-driven. And in an autobiography or anyone's life story, relationship tension occurs. Then the plot moves on. If it's a romantic suspense or mystery within a life story, such as true confession, true crime, or biography, usually twenty-four short chapters makes a book-length story.

A diary is written in first person as a journal or log, but a biography can be of you or your client. Even in a memoirs book or diary, you have to balance action with interaction between the heroine and the hero.

You can be the only person in your diary, but the action and interaction needs to be balanced with something out there in the external world—either forces of nature or another person—or the competition.

If you keep the competition out of your diary, put in the memories, actions, and warmth of the friends, including pets. If there are no other people, put in some force of spirit, some other push and pull, or tension, for balance with something outside yourself. This can be a job, school, a hobby, or what you choose as the force that pulls in an opposite direction existing with the force or person that pulls in your direction.

Try putting the relationship tension between the hero and heroine in the even-numbered chapters, and the mystery, historical events, or action plot events in the odd-numbered chapters.

In a 24-chapter-historical romance, this alternating action chapter, romantic tension chapter balances the plot smoothly. Most historical romance novels have 22-24 chapters. If you analyze the best-selling ones, you'll see that chapter one has an opening scene on the action side so you see what's happening.

The first action-oriented introductory chapter that shows us what's happening is followed by the second chapter on the romantic tension side showing us when and how the heroine meets the hero or has a re-union with the hero. In the second chapter, the writer takes the heroine somewhere in place or time. The heroine in the second chapter is defined. Either she's a 90's woman, or she's in her place in history or rebelling against it. You tell the story. If you're male, you'd use a hero.

Romantic life stories featuring genealogy combined with biography usually are 10-12 chapters long. Historical romances are twice that size at 22-24 chapters. The writer decides whether it's best to turn a biography into a historical romance or a life story into a mystery, suspense, action adventure, young adult novel, romance, or other genre.

If you are not fictionalizing genealogy or biography into a story, keep your time capsule book, database, or other media true to facts and historical records only. You might want to add your DNA testing records of relatives along with a family tree or other database or time capsule.

For those who want to turn factual biography into a novel, in turning a biography into a romance, the romantic tension side is about girl meeting hero in the first chapter. In the second chapter, the hero takes her somewhere in place, space, time, or state of mind. An oral history may be written as true life story in the form of a novel or play, skit, or anecdote of experience.

The oral history highlights a life experience within a time frame set in one or more locations with all the nuances of that place. It's basically a life story, but it can be transcribed with that certain something, including—charisma, liveliness, action, forward movement, drama, tension, and unique experiences, problems solved, and goals.

CHAPTER 23

▼

DIARIES AND DNA EQUAL EVIDENCE

What do diaries and DNA tests have in common? Both are handled as evidence.

Diaries, like deleted files hidden in the cache pits of computers, and DNA tests can be used as evidence. Diaries also hold the seeds of a story. You could write a novel or a screenplay from a diary. A diary also is a history. So preserve a diary as you would restore and preserve a valuable work of art from the past. Diaries are meant to be passed to future generations for a glimpse into a world that can be experienced by generations far into the future. Keep a file of dates listed in the diary and any objects that surrounded the diary from the same era.

A story with a central issue needed little explanation when my mom wrote in her diary in the style of a telegram: "October 25, 1926: First day of honeymoon. On train to Miami. Today I died." What central issues and themes tell a story in the diaries that cross your path?

Keep the dates and topics organized if you are working with restoring and preserving diaries. There should be a central issue or theme. How old was the person writing the diary? How many years did the individual keep the diary? What kind of objects were near the diary, packed together?

What kind of dust or other stains were on the diary—sawdust? Farm materials and plants? The first corsage from the senior prom? How about recipes, household hints, or how-to tips for hobbies? Was the diary or journal personal and inner-reflected, or geared toward outer events in the world? Were anecdotes about people and/or pets included, or was it about the feelings of the author of the diary?

Find out what other clues the mystery of the diary unfolds, from the lipstick or nail polish stain to the sawdust and coffee stains, or that faint smell of tobacco, industrial lint, or is it lavender, jasmine or farm dust and straw? Look inside the box in which the diary was packed. It's all evidence and clues waiting to be examined just like a mystery novel. A diary is a story, and everyone life deserves a novel, story, or biography and eventually, a place in a time capsule.

How to Restore a Diary

Make a book jacket for your diary to preserve and restore it. Use acid-free paper. Call a library or museum and ask for a brand or type of long-lasting acid-free paper and where you can buy some. Put a title and label on the dust jacket with the name of the diary's author and any dates, city, state, or country.

You can also speak to the art history department of most universities and find out what kind of paper is best to use for a book jacket to restore and preserve a diary. Treat it like a work of art. The same can be done for photo scrap books.

If torn, mend the diary. Your goal is to improve its condition. Apply a protective plastic wrapper to your valuable dust jacket. Give the diary a dust jacket in good condition. It should start to look more like a valuable book in good condition. If the diary is dingy and dirty, bleach it white on the edges. Put a plastic cover on the diary. The white pages of a diary without ink can be bleached with regular household bleach, but don't let the vapors of the bleach soak through to reach the ink because it will bleach out the writing.

Repair old diaries and turn them into heirlooms for families and valuable collectibles. The current price for repair of handwritten diaries and books is about $35 and up per book or bound diary, if you like to specialize in mending old dairies and family or personal books for a fee.

Some old diaries contain recipes and also served as personal, handwritten cookbooks containing recipes created by a particular family or family cook. These were valuable books preserved as if they were family scrapbooks, unlike the recipe databases in computers we have today. They are works of art, like an old tapestry embroidered with the story of a family's major turning points and events.

For more repair tips on bound diaries and books, I recommend *How to Wrap a Book*, Fannie Merit Farmer, Boston Cooking School.

How do you repair an old diary to make it more valuable to the heirs? You'll often find a bound diary that's torn in the seams. According to Barbara Gelink, of the Collector's Old Cookbooks Club, San Diego, to repair a book, you take a bottle of Book Saver Glue (or any other book-repairing or wood glue), and spread the glue along the binder.

Run the glue along the seam and edges. Use wax paper to keep the glue from getting where it shouldn't. Put a heavy glass bottle on the inside page to hold it down while the glue dries.

To remove tape, tags, or stains from a glossy cover, use lighter fluid or cleaning fluid (away from sparks, flames, or heat lamps). Dampen a cloth with nail polish remover if lighter fluid is too smelly and flammable for you.

Another way to remove something pasted on a plastic book cover is to use the finest grade sandpaper. Many books you'll find at goodwill will have adhesive price tags on the book. It's not usual to find diaries, even bound diaries in old book stores, but they show up in garage sales and in some antique stores and flea markets along with old photos.

To bleach the "discarded book stamp" that libraries often use, or any other rubber stamp mark, price, date, or seals on the pages of a book or on the edges, use regular bleach, like Clorox. It turns the rubber stamp mark white. The household bleach also turns the edges and pages of the book white as new.

To preserve a valuable dust jacket, a tattered jacket with tears along the edges needs extra firmness. A protective plastic wrapper can improve the condition of a book if it has a jacket cover.

To look for old diaries, or old family photos, look in garage sales, flea markets, and antique shops. Attend auctions and book fairs. Two recommended auction houses for rare cookbooks include Pacific Book Auction Galleries, 139 Townsend, #305, San Francisco, CA 94107, or Sotheby's, New York, 1334 York Ave., New York, NY 10021. Pacific Book Auction Galleries recently put a large cookbook collection up for an auction.

Hunt for diaries, old photos, and other old clues to family history in thrift shops and antique stores. Some diaries also combine cookbooks with personal histories and transcribed oral histories, but these are very rare. Genealogy also can have a person's collection of favorite recipes or anything else the person collected organized and archived along with family history and genetics records.

If you're into scrap booking with family history, photos, and recipes, for nostalgia, look for cookbooks printed by high school parent-teacher associations. Some old ones may be valuable, but even the one put out by the depression era San Diego High School Parent Teacher Association for the class of 1933-34 is only worth $10.

To find out-of-print and mail-order cookbooks, contact Charlotte F. Safir, 1349 Lexington Ave #9-B, New York, NY 10128-1513, (212) 534-7933. She specializes in hard-to-find cookbooks and children's books. Astor House Books, PO Box 1701 Williamsburg, VA 23187 (804) 220-0116, specializes in cookery and gastronomy. Amber Unicorn Books, specializes in rare cookbooks. They're at 2202 W. Charleston Blvd, #2, Las Vegas, NV 89102 (702) 384-5838.

Little Treasures (Joyce Klein) Cookery Books has British and American cookbooks and general stock. They're at 7517 W. Madison, Forest Park, IL 60130 (708) 488-1212. Send your wants because they have no catalog.

Cornucopia, run by Carol A. Greenberg, has cooking and food literature and domestic history, household management, herbs and kitchen gardens, hotels and restaurants, etiquette and manners, pastimes and amusements, and needlework old and rare books. They search for out-of-print books, and are interested in material from the 19th century through 1940. Write to: Little Treasures at PO Box 742, Woodbury, NY 11797, (516) 692-7024. Greenberg is always grateful for quotations on old, rare, and unusual materials in fine condition.

The Collector's Old Cookbooks Club has half their members in other states and half in San Diego County. They send a newsletter to each member after the monthly meeting.

You could specialize in being a diary restoration specialist and book finder for genealogy groups. Perhaps you want to deal in collectors' valuable diaries, largely first editions. Mostly diaries are only editions. Some people had them bound like a blank notebook, and wrote in them. So they look like first editions of books.

Are diaries worth as much as rare cookbooks? How much are the thousands of rare cookbooks worth today? A helpful guide is the Price Guide to Cookbooks & Recipe Leaflets, 1990, by Linda J. Dickinson, published by Collector Books, at PO Box 3009, Paducah, KY 42002-3009.

See Bibliography of American Cookery Books, 1742-1860. It's based on Waldo Lincoln's American Cookery Books 1742-1860, by Eleanor Lowenstein. Over 800 books and pamphlets are listed. Order from Oak Knoll Books & Press, 414 Delaware St., Newcastle, Delaware 19720 (302) 328-7232.

Louis & Clark Booksellers specialize in rare and out-of-print cookery, gastronomy, wine and beverages, baking, restaurants, domestic history, etiquette, and travel books. They're at 2402 Van Hise Avenue, Madison, WI 53705 (608) 231-6850.

Cookbooks and diaries are not that much distant from each other, although diaries of a famous person would have monetary value as would old cookbooks that are rare. Think of the events in the life story and history value of what real people's lives were like many decades or centuries ago. Any restored diaries would be valuable to descendants of anyone, and you can't put a price tag on these people's lives as expressed in diaries.

Get acid-free storage envelopes or boxes for diaries. Handle old diaries and books with gloves, and get rid of mildew safely without destroying the pages or fading the ink. It's here that a library can be helpful. Ask questions about storage from historical societies and libraries. Make copies of the diaries. Work with the photocopies when you decipher the writing.

Store old diaries in a dry, cool place where there are no bugs. Lining the storage place with plastic that's sealed will keep out vermin and bugs. Without moisture, you can keep out the mildew and mold. Store duplicates away from originals.

Call archivists and historians in your area and ask for their advice. Was something placed on a certain page, such as a dried rose, letter, or a special book mark? What meaning did it have? Date the diary. List the date it was begun and when it was ended, or look for clues for a time frame. List the geographic location of the events in the diary or the writer.

Of what kind of materials is the diary made? Is it improvised, created at low cost by the author? Or is it fancy and belonging to someone of wealth? What is the layout like? Does it show the education of the writer or anything personal? Was it a masculine almanac or calendar or a feminine expression of memorabilia and sentiment? Or was it written by a man writing poetry or letters and being romantic?

Can you tell the personality traits of the writer of that diary? What was the writing tool, a quill or a pencil? What's the handwriting like? Would a handwriting analyst know what to say about it? Does the handwriting and words express anger, joy, sadness, or what? Is it full of detail, or is the reader given the big picture? Why was the diary written? What is its central message? Do you see patterns, concepts or facts?

Transcribe the diary with your computer. What historical events influenced the writing of the diary? What language is it in or dialect? Study the historical meaning of the diary so you can get to know the writer across the chasms of decades or centuries.

Are there vital records such as deeds to real estate mentioned in the diary? What photographs are in the diary? Any artifacts mentioned or pressed between the pages? See whether you can relate to the diary author and find some type of context of the story from entry to entry. Are there any genealogical records or mention of family names, such as a great, great, great grandmother's maiden name? What about recorded events of inherited diseases or medical histories?

If the grandfather's dad went blind with glaucoma, it's has a genetic element that heirs should know about whether anyone inherited it or not. See what the diary unfolds that can be read as family history, world history, or used in the phenomics mode, to customize treatment or therapy based on genetics.

Treat the diary as a precious work of art, including the photos in there, if any. Don't touch the old right side up photos with bare fingers because the emulsion would quickly come off. Some genealogists specialize in working with old diaries, and novelists or screenwriters would probably be interested in a unique story. Restore a diary, and if a building such as a restaurant wants to hang old photos in the dining hall, let the photos be copies rather than the originals.

If you want to record the memories of relatives or friends, list 100-200 inspiring questions. Give the questions to relatives and friends. Create the questions with the goal of triggering recall of memories and experiences, events, and highlights of their life stories. Give them a diary and make a dust cover that's fancy. Put a plastic cover over the dust jacket.

Use a hardbound book with a question on each page. Or give several pages per question if the person might like to write for personal expression. You could ask the person if he or she wanted more or less pages. About 150 questions or pages are fine to last a lifetime of a minimum of one-page summarized life events or answers to the questions you ask.

The questions should be important to both you and the person who's going to write in the diary or journal. This is one way to create a genealogy diary gift that develops into a biography from a personal journal or diary.

Make sure there's enough pages and room to express what's required. Print the questions large and clear so you can elicit recorded responses to the questions. Leave room to attach photographs on acid-free, archival-quality paper.

If the person really won't write, create a tape recorded or video diary where the person can record his voice on tape, in a computer, or with a video camera on a tripod poised next to a desk to record the person talking without requiring anyone else in the room when recording.

Give the person a remote control device for the video camera to click on and off without walking toward the camera or tape recorder, or computer microphone. If the person will write at all, stick with the personal ambiance of a diary gift that someone can hold in any location or take traveling.

What questions to write? List thought-starters. Write questions about their childhood, values, dreams, and goals. Create a section for recording time as a mother or father, grandparent, how a child's name was chosen, how the marriages were arranged, or what things each child did to make a parent proud.

Put in blank pages and a place for each relative to put in the results of their DNA testing, fingerprints, print of hand or palm, or any other personal information, even medical notes or other memories in the memorabilia. Leave room for children's photos, small drawings, and meaningful relics. Use the kind of binding where someone can add pages.

File and archive the DNA test results, racial percentages tests, along with any other information from memory. Add an oral history, perhaps a pocket for a tape, or room for a transcript of an oral history to be added, such as a loose-leaf binding or similar binding so pages may be added to folders or plastic envelopes holding mementos.

Have a space for the first job and for a father's memories.

The genealogy thought-starter questions can come bound in a fancy, hard-bound loose-leaf that will last as pages are added. Or better yet to give each his or her own individual expression without the judgment of another relative, give a separate book to each family member so he or she can keep memories or traditions private until passed to heirs. You might want to create a separate book for each grandparent, parent, and other relative.

CHAPTER 24

▼

MAPPING YOUR PERSONAL ANTHROPOLOGY WITH GENETIC PROFILING

"Genealogy by genetics" is about historical records at the cellular level. And records profile how individual genes express themselves. Your genes also are time capsules. In the future, when space is colonized, instead of one species we may become many as we were in the beginning. It depends on whether we are isolated or actively trading genes. And over time, genes mutate, ever so slowly.

Source: Facts & Genes" <http://www.familytreeDNA.com/facts_genes.asp>.
Reprinted with Permission from:
Facts & Genes, from Family Tree DNA
Copyright 2002, Family Tree DNA),
November 21, 2002 Volume 1, Issue 5

=====================================

Family Tree DNA enjoys hearing about the results of your DNA testing. One customer wrote: "Just a word of thanks to you for the work you are doing on this project. The information I received today has already helped direct my genealogical work into a more focused and well-researched area, and has saved innumerable hours of work! Thanks for making this testing available, and for providing it in a financially accessible form. It's appreciated!"

Send comments, suggestions, tips, questions, and tell Family Tree DNA about your Surname Project to: editor@familytreedna.com.

Family Tree DNA is pleased to announce that the ANCESTRYbyDNA test is now available. The ANCESTRYbyDNA test was developed by DNAPrint Genomics, Inc., and is available through Family Tree DNA.

The ANCESTRYbyDNA test will measure a person's Personal Anthropology and their corresponding ancestral ethnic proportions. The result of the test is a report showing your percentages of each ethnic ancestry or major human population group. For example, your result could be 18% Native American, 70% European, and 12% African.

Perhaps you have wondered whether you have any Native American ancestry, or maybe you are just curious to find out more about yourself. The ANCESTRYbyDNA test will unlock the secrets to your ancestors contained in your DNA.

The ANCESTRYbyDNA test analyzes your DNA to determine which of the major human populations your ancestors belonged to, and what percentage you have inherited of these groups. These four geographical areas and the corresponding major human population groups are: Native American, East Asian, European, and sub-Saharan African.

This test, developed by DNAPrint Genomics, utilizes SNP's that are diagnostic of a person's continent of origin. SNP's are deep ancestral locations along the human genome, and have a different result when tested with different peoples.

To order the ANCESTRYbyDNA test, click on this link: http://www.familytreedna.com/products.html#dnaprintorder

Surnames

Are you wondering why the XYZ surname project has over 50 participants, and you only have 6 participants so far? Do you look at your Web site and correspondence, and wonder what is their secret to recruiting participants? The answer may be that they have a larger population of their surname from which to recruit participants. Your Surname Project may actually have a higher percentage of the surname participating than the project with over 50 participants.

It is common knowledge that Smith is the most frequent surname in the US. The chart below shows the 10 most frequent surnames in the US in the 1990 census. For each surname, the percentage represents the percentage of persons in the US with this surname, and the Rank is the ranking of the surname with 1 being the most frequent. For example, in the chart below, eight surnames are more frequent than Moore.

Surname	%	Rank
SMITH	1.006	1
JOHNSON	0.810	2
WILLIAMS	0.699	3
JONES	0.621	4
BROWN	0.621	5
DAVIS	0.480	6
MILLER	0.424	7
WILSON	0.339	8
MOORE	0.312	9
TAYLOR	0.311	10

Assume that a person started a Smith Surname project. There are over 2 million Smith's in the US, of which over 1 million would be males. This is quite a few people. If they signed up 50 people, they have only signed up a very small percentage of the Smith surname.

Compare this to the surname Mumma, which is .001 % of the population, and its Rank is 15,109. There is a much smaller pool of Mumma potential participants. If you look at the surname, Norin, its number is so small in the US 1990 census, that it does not even get a result when the 1990 US census Surname Frequency is searched.

You can find out what percentage of the US population holds your surname by going to the US Government census site at: http://www.census.gov/genealogy/www/freqnames.html

The site also covers the methodology that the Census Bureau used to come up with the percentages and rank for the surnames.

The US population on April 1, 2000 was 281,421,906 people. If you would like a rough idea of the males with your surname in the US, first search the site

<http://www.census.gov/genealogy/www/freqnames.html> to get the percentage
for your surname. Multiply that percentage times the population of the 2000
census. In their rough calculation, they will assume that 50% are males, so now
divide by 2. This is an estimate of the number of males with your surname. To
estimate the number of adult males, multiply by .7. The formula is:
Percentage*281,421,906/2*.7=adult males with surname

You can also find out how common your surname is in the UK at the site:
http://www.taliesin-arlein.net/names/search.php. There are 269,353 surnames in
the UK database, representing 54,412,638 people. This database is provided by
the Office of National Statistics of the UK, and gives an actual count of the number of persons for each surname. Their database is an extract of an Office of
National Statistics database, and provides a list of surnames in use in England,
Wales and the Isle of Mann in September 2002.

The US Census population database and the Office of National Statistics of
the UK database used different methodologies to come up with their results. Rare
surnames will not get a search result in the US census site, whereas they will in
the UK site, even if there are only a few persons with the same surname. Names
shared by fewer than five people have been excluded from the UK list.

Now that you have an idea of the size of your potential prospect pool, lets
assume that only 1/3 are interested in genealogy, so you now divide by 3. The
end result is a very rough approximation of the number of potential participants
available. If you are only using the Internet to find your participants, cut this
number in half for the US. Other countries have a smaller percentage of persons
on the Internet than the US.

As your first step, you have probably posted your project to as many sites and
mailing lists that are applicable and allow such postings. You have probably also
put up a web site, even if it is only one page. Most likely you have contacted all
those persons whom you had contact with in the past regarding genealogy.

Here are some suggestions to consider in order to make more people aware of
your project:

1. Consult the Directory of Family Associations. If there is a Family Associa-
    tion for your surname, contact them and offer to write an article for their
    publication about your project.

2. Register your web site with familysearch.org. Everyone searching on your
    surname at Familysearch.org will find your web site. You must first register

yourself with familysearch.org to be able to submit your website for consideration.

3. Visit your local Family History Center, and offer to show the Genealogy by Genetics video to the staff and patrons. This might not find you any participants, but if every Group Administrator takes an hour to do this, then all the Surname Projects might find participants.

4. Review your web site. It needs to be easy to understand for those not familiar with DNA testing, and clearly present the benefits to the participant. What will they gain from participating? How will it help them in their research? What might the results tell them?

5. Find out if there are any genealogy clubs or organizations in your area, and volunteer to show the video, and answer questions.

DNA testing for genealogy is a new field, and we are all pioneers. Most likely you have learned a lot about the field as a result of your testing. Those of us who have learned about DNA testing and how to interpret the results are aware of the benefits and how the testing can assist us with our genealogy research. Convergence explains why a haplotype will match others with a different surname. DNA testing for genealogy is not a substitute for genealogy research, but is instead a companion.

The majority of those interested in Family History research may not yet be fully aware of Genetic Genealogy. If you volunteer an hour to help your fellow genealogists understand this new tool, and help more people become knowledgeable, all of us will benefit as we seek participants for our testing. Look for social histories of the ethnic group you're researching.

When you're working with DNA, you can look for historical medical records. Only a few are open to the public. You might try the microfilmed collections at The Family History Library in Salt Lake City, UT, or rent one of the microfilms from any of the worldwide family history centers, usually found in a genealogical library in various cities around the world.

Look at records from the Eugenics Record Office (ERO) that operated from 1910 to 1944. The purpose of that office was to study human genetics in order to reduce inheritable genetic disorders. You can look over the 520 rolls of microfilm. Visit the Family History Library Catalog and look under United States—Medical Records—Eugenics.

Look up the state you want, and look under Medical Records. You might want to read up on the controversial Eugenics movement. Think about how DNA testing today differs in that the test results are used today either to find relatives, ancestors, or tailor individual therapies for individual genetic makeups—phenomics. How times have changed. Or have they? What do you see in your own DNA and family history research?

The Family History Library in Salt Lake City also has some historical hospital records. One example is the Northwestern Memorial Hospital record, Chicago in the Family History Library, dated 1896-1933. Perhaps one of your relatives is in those files. That's one other way of finding a maiden name from the days when many people were never given a birth certificate because they were born at home and never registered. That's what happened with my mom, born in 1904 at home in NY state.

Understanding your Results: Ethnic Origin

Whether you are just starting a Surname Project, or ordered a test for yourself to learn about DNA testing for genealogy, everyone experiences the situation of receiving the first test result, and what now? You have one test result, and what do you do with a string of 12 or 25 numbers? Can they tell you anything? Where can you find the information you need?

In the situation of the one or first test result, most likely you will not find others to whom you are related. The odds of a random match to some one to whom you are related when you are the first of your surname to test is slim to none. Instead, you might find some clues to your ethnic origin.

To find clues about your ethnic origin, Log into FamilyTreeDna.com, and at your Personal Page click on Recent Ethnic Origins to search this data base. The results show others whom you match, or who are a near match, and their ancestor's ethnic origin.

The information on their ethnic origin is provided by each person tested (testee). The information provided for ethnic origin is only as accurate as the knowledge held by the testee regarding their ancestors. Testees are instructed to answer unknown for ethnic origin when their ancestor's origin is not known, or not certain. Sometimes the origin the testees provided is incorrect.

Incorrect origins provided by testees may lead to search results that do not seem logical. For example: Assume your ancestors are from England, but your search results show the ethnic origin of your matches as England, France, AND one match shows an origin of Native American. Does that mean that your ancestor's relatives may have lived in England and France? Yes. Does it mean that your ancestor was also a Native American? No. It means that a settler in America had a child with a Native American woman, the child was brought up as a Native American, and that, over time, the family has "forgotten" the European ancestor, and believes their ancestry to be Native American.

During the span of generations people tend to move, as do borders, so nationality or ethnicity becomes subjective. For example, testees may enter Germany for ethnic origin, because the land of their ancestors is in Germany today, but the land had been held by Denmark for many centuries.

Your search should return at least one match, namely yourself. If your results show 3 matches from Ireland and 1 from Scotland, and you have reported to us that your ancestors came from Scotland, then you are the Scotland result. The other 3 matches are either from the Family Tree DNA database or from the databases Family Tree DNA have been supplied by the University of Arizona.

To see how your ethnic origin is recorded in our database, click on the link titled Update Contact Information. You can also update your paternal and maternal ethnic origin on this Update Contact Information page.

Exact matches show people who are the closest to you genetically. The Ethnic origin shows where they have been reported to have lived. Since many persons migrated since the beginning of time, you will typically see matches in more than one country.

For information purposes, the Recent Ethnic Origin search also displays results for those who are not exact matches, but are 'near matches'. A near match is either one step or two steps from your result. An exact match is 12/12 or 25/25. A one step match is 11/12 or 24/25. A two step match is 10/12 or 23/25. The value of the near matches is to see where those who may be related migrated over time.

Other databases available that you can search are:

European: http://ystr.charite.de/index_gr.html

US: http://www.ystr.org/usa/

In some cases you will not find any results. This is because only a very small percentage of the world population has been tested and is in the databases. The Ystr databases, plus the FamilyTreeDNA Recent Ethnic Origin database together hold about 21,000 test results. Every day more results are added, and it is only a matter of time before you will have some matches. Your test with Family Tree DNA includes access to our databases for matching.

If you do not find any results in the two YSTR databases shown above, try entering your result, and then eliminating a marker, and do this until you have a smaller set of markers that results in some matches. This might provide some clues regarding where your markers have occurred geographically.

The value of DNA testing comes from comparing your results to others. If you have started a Surname Project, you will most likely have results from others soon. If you only tested yourself, you may want to consider either using DNA testing to solve one of your Family History questions, or starting a Surname Project.

Haplotypes: Convergence

A Haplotype is the 12 Marker result from testing the Y chromosome. Some Haplotypes are common, with a high frequency of occurrence and some Haplotypes are rare, with a low frequency of occurrence.

Many people have common Haplotypes, which means that they would expect to find matches to those who do not have their surname. This occurs because we were all at one point related. As the different branches of the Adam+Eve tree evolved throughout time, mutations occurred, forming different Haplotypes. Thousands of years later, you have many different Haplotypes. Due to these mutations, you could have two branches that mutate to an identical Haplotype. This is called convergence.

If your Haplotype matches an individual with a different surname, and your genealogy research shows no evidence of an extra-marital event or adoption, your match may be the result of Convergence.

The example below shows convergence between the ABC surname and the XYZ surname, using just 3 markers to keep the example simple. Notice how the mutations over time bring two different Family Lines to the point that they match.

Time	ABC	XYZ
1000 A.D.	12 24 15	14 25 13
1200	13 24 15	14 25 13
1400	13 24 15	14 25 14
1600	13 24 15	14 24 14
1800	13 24 15	13 24 14
2000	13 24 14	13 24 14

Convergence explains why a haplotype will match others with a different surname. DNA testing for genealogy is not a substitute for genealogy research, but is instead a companion. Results that match must be considered in light of the genealogy research. If you match someone with a different surname, most likely there wasn't an adoption or extra marital event, and your match may be the result of convergence.

Case Studies in Genetic Genealogy

In each issue of the Newsletter, Family Tree DNA looks at what Genetic Genealogy will do for your Family History research. This article is a continuation of the topic, with situations, called "*Case Studies*", followed by a recommendation. The objective of the case studies is to present different situations you may encounter in your family history research, and how DNA testing can be applied.

Case Study

From November 21, 2002 Volume 1, Issue 5, *Family Tree DNA Newsletter*, "I have participated in a Surname Project, and had quite surprising results. All the other Lines of my surname are related, except my Line. We have all traced our ancestors to England. Not only is my line not related, but also my ethnic origin is Eastern European. What do I do now?"

Recommendation

From November 21, 2002 Volume 1, Issue 5, *Family Tree DNA Newsletter*, "I am sure you were quite surprised, and perhaps disappointed. The first step is to validate the result for your Line or family tree. Since only one person was tested for your Line, we recommend testing additional males from each branch on your tree, to see if they all match each other. If they end up matching, your result is probably due to an extra marital event, an adoption, or a name or spelling change.

"In reviewing the surnames of Eastern Europe, your surname is pronounced as the surname in England, only the spelling is different. A review of your Family History shows that the research and documentation for the time period 1800-1850 is quite sparse. Many more records are available in England for this time period, including parish registers and wills. I would suggest that more family history research might shed some light on the situation."

*   *   *   *

Managing a Genetic Genealogy Project: Participants with Poor Documentation

Occasionally you might run across a willing participant for your Surname Project who has a poorly documented family tree, perhaps even built entirely out of the International Genealogical Index (IGI) by matching surnames. Your dilemma is that the prospective participant appears to be from a Line you haven't tested yet, but without better research you can't be sure. What comes first, the testing or the research?

This is a complex issue. If you turn away the participant and suggest that they do more research, they may become discouraged, and never return. If the participant tests, and gets unexpected results, they may become an unhappy participant.

One solution is to fill in the gaps of their research. You may not have the time to take this step. A better solution may be to communicate the situation to the participant, and let them make the decision to test now with the possibility of unexpected results, and also encourage them to do further research.

Perhaps from your research experience, you may be able to suggest to the participant specific sources for them to investigate. Most likely, they want to do more research, and just need some guidance and direction.

It will be a win-win for both the Surname Project and the participant if you are able to achieve both additional research on their part, and their participation.

Spot Light: Witt—Whitt Surname Project

Objective: Prove or disprove the genealogy research of the Witt/Whitt Line from Old Virginia

There are three identified Lines or families of the Witt/Whitt surname in the US. One family Line that today spells their surname as both Witt and Whitt begins with German immigrants in both South Carolina and Pennsylvania in the early 1700s. A second family Line that today spells their surname as Witt and DeWitt, began in New England around 1640 with an English immigrant by the name of John Witt.

The third family Line that today spells their surname as Witt and Whitt began with an individual named John Witt or Whitt, who first appears in early records in 1670 in colonial Virginia. The records relating to John show the spelling of his name both as Witt and as Whitt. It was from John Witt-Whitt that the Witt-Whitt Family of Old Virginia began.

Early Virginia records, John Witt-Whitt was the father of at least four sons, John Witt II, William Witt, Edward Whitt, and Richard Whitt. The participants in the Witt-Whitt Surname Project are all documented descendants of the 4 identified sons of John Witt-Whitt: John Witt II, William Witt, Edward Whitt, and Richard Whitt. For each of these sons, at least two documented male descendants participated in the Project. All participants took the 25 marker test.

The results for this Surname Project are that the majority of participants matched 25/25, and a few matched 24/25. Therefore, the Project has confirmed the genealogy research, and shows that the participants are related and have a common ancestor.

When combined with surviving colonial Virginia records for the surnames Witt and Whitt, the Witt-Whitt DNA study determined John Witt II, William Witt, Edward Whitt, and Richard Whitt were brothers and their father was the immigrant John Witt-Whitt of Charles City County, Virginia. The Witt-Whitt DNA Surname Project also identified the common ancestor of these four men was from England, or possibly Scotland.

A DNA baseline for the Witt-Whitt family of Old Virginia has now been established. Other descendants who have incomplete records, or where records no longer exist and preclude the determination of a family's origin, may take the 25 marker DNA test to determine if they are related to the Witt-Whitt Line from old Virginia.

If other descendants find that they match, they can contact one of the participants in the baseline study to share the Witt-Whitt family of Old Virginia ancestral history for their family line. The next phase of the Witt-Whitt surname project is to identify the county in England or Scotland from which John Witt-Whitt originated.

If you're interested in receiving Facts & Genes newsletter, feel free to contact the editor at Family Tree DNA with your comments, feedback, questions to be addressed, as well as suggestions for future articles. If you would like your Surname Project featured in their *Spotlight* column in a future issue, please send an email telling them about your project. If you are a Project Manager and can help others with tips or suggestions, please contact the editor:
editor@familytreedna.com

Reprint Policy:

Family Tree DNA encourages the circulation of "Facts & Genes" by newsletters and lists providing that you credit the author, include their copyright information (Copyright 2002, Family Tree DNA), and cite "Facts & Genes" (http://www.familytreeDNA.com/facts_genes.asp) as the source.
November 21, 2002 Volume 1, Issue 5

Reprinted with permission, from November 21, 2002 Volume 1, Issue 5, Family Tree DNA Newsletter.

CHAPTER 25

▼

HAPLOGROUPS AND MARKERS: WHAT'S A RECESSIVE GENE?

Sometimes a recessive gene is referred to as a form of a gene called a recessive allele. The recessive allele will not express itself if combined with a dominant allele. The recessive allele is expressed by a lower-case letter. Some traits may be caused by having two recessive alleles.

Markers

How many genetic markers can tell us something for various pairings between groups of people? Markers often are great for telling multiple groups apart. For example, one marker in particular can tell Africans apart from all the other groups, but that marker can't tell Europeans from East Asians. Scientists also will look at male Y chromosomes to study various markers.

Markers are different for various ethnic groups. However, there also is some overlap as peoples become mixed. A geneticist can tell the percentages of various races by looking at the markers, even though we have come to accept there really is no such thing as a particular race because of the diversity between peoples in any one race. However, you can still look at genetic markers to see various ethnic group traits for people who have been separated for thousands of years.

African-European	54 markers
African-East Asian	50 markers
African-Native American	50 markers
European-East Asian	45 markers
European-Native American	41 markers
East Asian-Native American	24 markers

What is a Haplogroup? How is it different from a Haplotype?

Your matrilineal or female ancestors inherit the same mtDNA sequences which form a haplogroup. Look at female lineages starting with your mtDNA haplogroup today. It will be the same haplogroup letter as your common ancestor with the same haplogroup letter who lived 21,000 years ago. You are looking at a connection from a single female ancestor to all your direct female line ancestors today.

The sequences within the haplogroup may be slightly different because of the slow mutation rate of the mtDNA, but the haplogroup will be the same. And in some cases, the sequences will be similar to your ancient ancestors. While in other cases, the mutation rate may have changed your mtDNA just a little over that long span of time.

What's a Haplotype?
Let's look at the female lineages, the mitochondrial DNA clues called mtDNA for short.

Individual mitochondrial DNA called for short, mtDNA sequences, is grouped into haplotypes. A haplotype defines a series of special mutations. The mutations when lumped together are called haplogroups. Each haplogroup contains a set of haplotypes descended from the same one common ancestor. How many haplogroups of mtDNA are there? According to Bryan Syke's book, *The Seven Daughters of Eve,* at least 35 mtDNA haplogroups represented by a letter of the alphabet are listed in one of the illustrated tables. Matrilineal (female ancestral) lineages contain the mitochondria.

You can look at ancient ancestry by tracing the mtDNA lines. Some mtDNA letters belong mostly to Africa, while others belong to East Asia. Some are specific to South West Asia (India) and others are found in central and west Eurasia, which includes Europe and the Middle East. Five different mtDNA haplogroups are found in the New World—the Americas, such as ABCD and X, but the differences between the European and Middle Eastern X and the X among some Native American peoples show that they have been separated for thousands of years.

For example, in the book *Mapping Human History*, by Steve Olson, a rare and unusual haplogroup, X showed up among the Algonquian-speaking Native Americans living around the Great Lakes. It also is present in small amounts in the Lakota and Sioux. Previously mtDNA haplogroup X had been found in Finland, and in Italian, Greek, and Druze (Israel and Lebanon) peoples. Haplogroup X so far has not been found in East Asians. How did it get to the New World?

The Native American X mtDNA differed very much from the European haplogroup X to be separated by only one or two thousand years. It had to have come to the New World tens of thousands of years ago. Scientist Douglas Wallace of the Center for Molecular Medicine at Emory University in Atlanta is one of the world's leading experts on mitochondrial genetics. So when he studied two skeletons that lived in the 1300s in Illinois among the Native Americans of that time, he found the skeletons contained traces of haplogroup X.

How did he know it wasn't from mixture with a European? It was the time divergence between the European and the Native American X haplogroup that gave the answer. The haplogroup X in North America had been there for more than 10,000 years. It wasn't a "modern" European who lived in Illinois in the 14th century.

Again, you might ask, perhaps it was a Viking from Finland since X is found in Finland? The tests showed this type of X differs from the European X by mutations that reveal the X that lived in America really had been there more than 10,000 years. So it could have come more than 10,000 years ago from anywhere—central Asia, Siberia. No X haplogroups are in Siberia today as far as one can tell.

Then again, not everyone has been tested there. However, you have to draw the line somewhere, and the differences between the old world and new world X haplogroup were great as if they had been separated more than 10,000 years. It's easy to imagine someone in north central Asia could have joined up with a group of people such as hunters and traveled with them over the Bering Strait while it was still a land bridge more than 12,000 years ago.

What's a Haplogroup?

A group of related <u>haplotypes</u> make up a haplogroup. Haplogroups are studied especially when referring to <u>mitochondrial DNA</u> and Y-chromosomes. If a set of haplotypes are placed into a tree determined by the minimum number of mutations that separate them, the main branches of that tree are haplogroups. Each haplogroup in theory contains haplotypes that are all descended from a single founding individual.

Haplotypes from other regions of the genome are not studied as much because they may not always group together. Recombination makes ancestor-descendant relationships not as specific to see. You have to look for connections when you study haplotypes.

Examples: The vast majority of Native Americans belong to one of four mtDNA haplogroups: A, B, C, and D, but a few Native Americans also belong to haplogroup X. Haplogroup X is found at a low percentage in Europe, but the differences between the European haplogroup X and the Native American haplogroup X show that they separated more than 10,000 years ago.

It's more likely that someone with haplogroup X mtDNA from Southern Siberia, the Caucasus, or central Asia joined a group of hunters headed north and east more than 12,000 years ago when there was a land bridge over the Bering Strait, and settled in what is now called the North American continent.

* What's an allele?

For lots of definitions of these terms, also see the Web site: library.thinkquest.org/18258/noframes/def-allele.htm. Or See:
www.apnet.com/inscight/08271998/allele1.htm

An allele is a form of a gene. Alleles are located at the same position (locus) on homologous chromosomes and are separated from each other during meiosis. An allele is what is actually within a region of the chromosome, and is found within a gene. *An allele is any of two or more alternative forms of a gene occupying the same chromosomal locus; such as that which determines flower petal color in peas.*

* What's a Haplogroup?

Definition: A bunch of haplotypes make up a haplogroup. The term is used usually when referring to female lineages and mitochondrial DNA or mtDNA. You might call a form of a gene an allele. An allele is an alternative form of a genetic locus. A single allele for each locus is inherited from each parent (e.g., at a locus for eye color the allele might result in blue or brown eyes).

So when a group of alleles on a single chromosome are linked together and usually inherited as a unit, these genes make up a haplogroup. Haplotypes are particularly stable in <u>mitochondrial DNA</u> and on the Y-chromosome, because they are not subject to recombination.

Analyses of mtDNA and Y-chromosome variation usually focus on the haplotype or <u>haplogroup</u> level, rather than comparing exact base pair sequences. In this case, haplotypes are defined on the basis of particular mutations shared by various individual DNA lineages.

Examples: One study of Finnish Y-chromosome variation found that 40% belonged to one of two different haplotypes, which in turn each belonged to different haplogroups and were probably introduced by different founding populations.

*Genome. A person's genome is one set of his (or her) genes. The human genes, which control a cell's structure, operation, and division, are located in the cell's nucleus. The full human genome (estimated at 50,000 to 100,000 genes) is present in every cell-nucleus. Many genes are inactive in cells that have some specialized functions. Many cells are differentiated to perform certain functions only.

*Genes and Chromosomes. Genes are composed of segments of DNA. In normal cell-nuclei, the DNA is distributed among 46 chromosomes (23 inherited at conception from a person's dad and 23 from mom). Each chromosome consists of one very long strand of DNA and numerous proteins.

The proteins are needed to manage the long DNA molecule. The longest chromosomes each support thousands of genes. Every time a cell divides, the cell must duplicate the 46 chromosomes. Every cell must distribute one copy of each chromosome to the two new cells. When cells stop dividing, that's the end of them and the organism.

*The DNA Code. The DNA of each chromosome is composed of units— "nucleotides" of four different types (A, T, G, C). These nucleotides are linked to each other in linear fashion. The necessary sequence of the four types of nucleotides produces the "code" which first determines the function of each particular gene. Then the sequence identifies the gene's start-point and stop-point along the DNA strand. Finally, the sequence allows specified regulatory functions. The code of the human genome consists of more than a billion nucleotides.

The Mitochondrial DNA (mtDNA). Mitochondria are needed for energy in the cell. The mitochondria are inherited from the mother. When tracing ancient and modern ancestry, geneticists look at female lineages or mtDNA. Your mtDNA is passed from mother to daughter over tens of thousands of years with few changes.

MtDNA mutates slowly during thousands of years of migrations of people across the globe. Men inherit their mtDNA from their mothers, but pass on their Y chromosomes to their sons. Women pass their mtDNA to their daughters. On very rare occasions, a few women may inherit some mtDNA from their fathers, but almost all women inherit their mtDNA from their mothers.

What mtDNA Does Not Do: It is not junk DNA. Often DNA produces more copies than it needs to function. Sometimes this is called junk DNA. MtDNA is necessary for providing energy to the cell. Outside the nucleus, human cells also have some "foreign" DNA located in structures called the mitochondria. This small and separate set of DNA does not participate in the 46 human chromosomes.

The mitochondrial DNA (mtDNA) really is not part of "the genomic DNA." According to the "out of Africa" theory that's widely held in acceptance by most scientists, all the mtDNA in the world today came from a single woman called Mitochondrial Eve who had two daughters who survived to create a line of females that expanded all over the world. Similar histories are noted for the male line using the Y chromosome. According to the book, *Mapping Human History*, by Steve Olson, (page 56) "All non-Africans descend from Africans who left the continent within the past 100,000 years."

According to a number of the latest videos on whether people took the northern or southern route out of Africa, the southern route is most favored. According to the book, *Archaeogenetics*, the flow of people varied between then and now. Today, most scientists theorize that since the north route out of Africa most likely was blocked by an Ice Age that created a dry desert in the Middle East, those leaving Africa headed toward Yemen and then along a southern route to India, Malaysia, and finally Australia.

Only when the climate changed and the Fertile Crescent of the Middle East opened up, did people expand back from India toward where the rivers met, the Middle East, such as what today are Iraq and Iran, and the Levant, reaching the coast, and from there north to what today is Europe. About 21,000 years ago, a new Ice Age began, and people who moved up from the Middle East into Europe found refuge in only a few places such as Southwest France, Northern Spain facing the Mediterranean and the Pyrenees, the Balkans and Ukraine, until the last Ice Age ended about 12,500 years ago. Then populations expanded across Europe from Spain to the Urals.

By that time, the Far East had been populated for a long time, and Central Asia was the newest land to be seen. Then by 9,000 years ago, farmers from the Levant and Anatolia moved into Europe and introduced the idea of farming so that about 80 percent of Europeans today consist of the old Paleolithic hunters and about 20 to 26 percent from the more recent arrivals from the Middle East, the cereal belt grain farmers of the Neolithic era that started about 10,000 years ago in the Levant and Fertile Crescent of the Middle East.

Genealogy, history, folklore, oral history, memoirs writing, diary journaling, demography, anthropology, and archaeology are in the midst of a molecular revolution. Has archaeology become archaeogenetics? Actually, molecular genetics biotechnology is one more *tool* in the hands of the genealogist, historian, archaeologist, folklorist, prosopographer, onomasticist, demographer, videographer, anthropologist, or family historian. And that tool, molecular genetics, is used to untangle distantly ancestral as well as recent family roots.

Now you have computer technology and Web databases to research family ties. You have molecular genetics biotechnology—DNA testing, bioinformatics, and beyond.

From ancestry by DNA to racial percentages by markers and phenomics, experts can customize medicine or therapy to an individual's genes.

You can take a paternity test. Or find out whether you're related to a distant cousin you never met. Or you can study DNA as legal evidence. Your genes are used for matching bone marrow donors to recipients. The molecular revolution is enhancing research. From community colleges where students earn one or two-year certificates in biotechnology to perform DNA testing and bioinformatics processing on computers to the PhDs who work in research labs and universities, the molecular revolution has now joined science to history. How much do you want to know about your genome?

CHAPTER 26

▼

CANCER GENETICS
NETWORK

Reprinted with permission from the Cancer Genetics Network (CGN).

Have a Personal or Family History of Cancer? Consider Joining the Cancer Genetics Network

    *The Cancer Genetics Network (CGN) seeks individuals with a personal or family history of **cancer** who may be interested in participating in studies about inherited susceptibility to cancer. Nearly 8,500 individuals have enrolled in this unique program.*

    The Network is becoming an important vehicle to conduct studies that will provide much-needed <u>clinical</u> information to help individuals who may be at increased risk for cancer because of a personal or family history of the disease.

    Eight U.S. centers, funded by the National Cancer Institute (NCI), joined forces 3 years ago to establish a national resource to support investigations into the <u>genetic</u> basis of cancer susceptibility. Together, the centers are working to make possible research that a single institution may not be able to accomplish because of insufficient numbers of participants, or the time needed to recruit them.

"The idea is to have a pool of interested individuals readily available so that important research questions can be answered, and studies can progress without unnecessary delay," said Deborah Winn, PhD, acting associate director of NCI's Epidemiology and Genetics Research Program (EGRP), Division of Cancer Control and Population Sciences (DCCPS). Participants may be invited to be part of specific studies, depending on the research requirements, and may choose to participate on a study-by-study basis.

Questions in Search of Answers

The Network's emphasis is on supporting research that brings the tremendous knowledge about genetics gained from laboratory research to bear on improving prevention, screening, diagnosis, and treatment of cancer in humans. "A wealth of new information on genetics has emerged over the past decade, and the challenge now is to find out how to make these findings meaningful in clinical practice and for public health programs," said Dr. Winn.
Some of the pressing questions that the Network aims to address are:

- How common are the genetic changes (alterations) that cause cancer in different groups?

- What determines whether someone with a genetic change gets cancer?

- What environmental exposures interact with genetic susceptibility to cause cancer?

- How can genetic discoveries be translated into better ways to prevent and treat cancer?

- What ethical, psychological, social, and family issues affect healthy individuals and their families who carry cancer susceptibility gene alterations?

"The Cancer Genetics Network is uniquely suited to support research centered on the study of key interactions between external environmental exposures and inherited susceptibility factors for cancer," said Joellen Schildkraut, Ph.D., of Duke University Medical Center, Durham, NC, and chair of the Network's Steering Committee. "This research can lead to the design of timely interventions, such as behavior modifications and chemoprevention strategies that prevent cancer or halt its progression."

Being Part of the Network

The Network offers individuals an opportunity to keep up to date on cancer genetics and potentially to participate in studies. All Network centers are enrolling eligible participants, and are especially interested in recruiting minorities, among whom membership lags. "We want all groups to be able to take advantage of this opportunity and to benefit from the studies," said Dr. Winn.

Participants provide information about their personal and family medical histories, which is entered into a central database that is operated by an informatics group. Presently, information on more than 134,000 family members is in the database. All information is kept private and is protected by the latest communications technology safeguards.

Network researchers and their centers have longstanding experience in working with individuals and families at increased risk for cancer, and will confidentially consult with individuals who are interested in joining. "The ultimate aim is to prevent cancer, and our best hope for developing effective cancer prevention programs lies in the early identification of high-risk populations and individuals at high risk," said Dr. Schildkraut.

Pilot Studies Under Way

Although still a young program, the Network is conducting a variety of pilot studies. It also has begun to work with other research groups, and welcomes new opportunities to cooperate on important research. Some of the pilot studies under way, or slated to begin soon, are to:

- Test the value of screening for ovarian cancer among women at high risk for the disease using a blood test for CA–125 (a chemical found in the blood) and transvaginal ultrasound;

- Search for novel regions on genes associated with susceptibility to colon cancer among siblings who have a history of the disease;

- Obtain and characterize biological specimens from families who have a history of onset of prostate cancer at an early age;

- Study genetic and environmental factors that may modify risk for developing breast or ovarian cancer among women who are carriers of <u>BRCA1</u> and <u>BRCA2</u> gene alterations; and

- Compare statistical models for estimating the likelihood that a woman has a BRCA1 or BRCA2 gene alteration based on her family history.

How to Contact the Network

Individuals may contact one of the Network centers to discuss enrollment. It is not necessary to live near a center in order to join. Some centers have hospital affiliates through which one can enroll, and much of the contact can be by telephone, mail, or e-mail. More information about the Network is available on NCI's Web site: <u>http://epi.grants.cancer.gov/CGN</u> on the Internet.

Carolina-Georgia Cancer Genetics Network Center

Institutions:	Duke University Medical Center, Durham, NC, in collaboration with the University of North Carolina at Chapel Hill, NC, and Emory University, Atlanta, GA
Principal investigator:	Joellen Schildkraut, Ph.D., Duke University Medical Center
CGN Web site:	http://cancer.duke.edu/CGN
Institution:	Duke University Medical Center
Contact:	Sydnee Steadman
Telephone:	888–681–4762 (toll free)
E-mail:	stead006@mc.duke.edu
Institution:	University of North Carolina at Chapel Hill
Contact:	Cindy Smith
Telephone:	877–692–6960 (toll free)
E-mail:	cesmith@med.unc.edu

Institution:	Emory University
Contact:	Lisa Susswein
Telephone:	800–366–1502 (toll free)
E-mail:	lrs@rw.ped.emory.edu

Georgetown University Medical Center's Cancer Genetics Network

Institution:	Georgetown University Lombardi Cancer Center, Washington, DC
Principal investigator:	Claudine Isaacs, M.D.
CGN Web site:	http://lombardi.georgetown.edu/research/areas/cancercontrol/cgn
Contact:	Camille Corio
Telephone:	202–687–8070
E-mail:	corioc@georgetown.edu

Mid-Atlantic Cancer Genetics Network Center

Institutions:	Johns Hopkins University, Baltimore, MD, in collaboration with the Greater Baltimore Medical Center
Principal investigator:	Constance Griffin, M.D., Johns Hopkins University
CGN Web site:	http://www.macgn.org
Contact:	CGN Staff
Telephone:	877–880–6188 (toll free)

Northwest Cancer Genetics Network

Institutions:	Fred Hutchinson Cancer Research Center, Seattle, WA, in collaboration with the University of Washington School of Medicine, Seattle
Principal investigator:	John D. Potter, M.D., Ph.D., Fred Hutchinson Cancer Research Center
CGN Web site:	http://www.fhcrc.org/science/phs/cgn
Contact:	CGN Staff
Telephone:	800–616–8347 (toll free)

Rocky Mountain Cancer Genetics Coalition

Institutions:	University of Utah, Salt Lake City, UT, in collaboration with the University of Colorado, Aurora, CO, and University of New Mexico, Albuquerque, NM
Principal investigator:	Geraldine Mineau, Ph.D., University of Utah, Salt Lake City
CGN Web site:	http://www.hci.utah.edu/cgn
Institution:	University of Utah
Contact:	Debra Dutson
Telephone:	877–585–0473 (toll free)
E-mail:	ddutson@hci.utah.edu
Institution:	University of New Mexico
Contact:	Lloryn Swan
Telephone:	505–272–5659
E-mail:	swan@nmtr.unm.edu
Institution:	University of Colorado
Contact:	Theresa Mickiewicz

Telephone: 877–700–0697 (toll free)

E-mail: theresa.mickiewicz@uchsc.edu

Texas Cancer Genetics Consortium

Institutions: University of Texas M.D. Anderson Cancer Center, Houston, TX, in
collaboration with the University of Texas Health Science Center at
San Antonio, University of Texas Southwestern Medical Center at Dallas, and Baylor College of Medicine, Houston

Principal
investigator: Louise C. Strong, M.D., M.D. Anderson Cancer Center

CGN Web
site: http://texas.cgnweb.org

Telephone: 877–900–8894 (toll free)

Institution: Baylor College of Medicine

Contact: Sharon Plon, M.D.

Telephone: 713–770–4251

E-mail: splon@bcm.tmc.edu

Institution: University of Texas Southwestern Medical Center

Contact: Gail Tomlinson, M.D.

Telephone: 214–648–4907

E-mail: tomlinson@simmons.swmed.edu

Institution: University of Texas Health Sciences Center

Contact: Susan Naylor, M.D.

Telephone: 210–567–3842

E-mail: naylor@uthscsa.edu

Institution: University of Texas M.D. Anderson Cancer Center

Contact: Louise C. Strong, M.D.

Telephone: 713–792–7555

E-mail: lstrong@mdanderson.org

UCI-UCSD Cancer Genetics Network Center

Institutions: University of California at Irvine and University of California at San Diego

Principal
investigator: Hoda Anton-Culver, Ph.D., UC Irvine

Contact: CGN Staff

Telephone: 949–824–7401 (collect calls accepted)

University of Pennsylvania Cancer Genetics Network

Institution: University of Pennsylvania Cancer Center, Philadelphia, PA

Principal
investigator: Barbara Weber, M.D.

Contact: Rhonda Kitlas

Telephone: 888–666–6002 (toll free)

E-mail: kitlasr@mail.med.upenn.edu

Informatics Infrastructure

The CGN also has an Informatics and Information Technology Group to meet its information exchange and data management and statistical needs. The participating institutions and principal investigators are the University of California at Irvine, with Hoda Anton-Culver, Ph.D.; Massachusetts General Hospital, Boston, MA, with Dianne M. Finkelstein, Ph.D.; and Yale University, New Haven, CT, with Prakash M. Nadkarni, Ph.D.

✳   ✳   ✳   ✳

Sources of National Cancer Institute Information

Cancer Information Service

Toll-free: 1–800–4–CANCER (1–800–422–6237)

TTY (for deaf and hard of hearing callers): 1–800–332–8615

NCI Online

Internet

Use http://cancer.gov to reach NCI's Web site.

LiveHelp

Cancer Information Specialists offer online assistance through the *LiveHelp* link on the NCI's Web site.

Have a Personal or Family History of Cancer? Consider Joining the Cancer Genetics Network

The text of National Cancer Institute (NCI) material is in the public domain when the content was written by a government employee. Such content is not subject to copyright restrictions. One does not need special permission to reproduce or translate written text created by NCI staff. However, we would appreciate a credit line and a copy of any translated material.

Likewise, permission is not needed to link to NCI Web sites. If you wish to use material on NCI Web sites, we strongly suggest linking directly to that information to be sure that you have the most up-to-date version.

*   *   *   *

What is the Consumer's Role in Nutritional Genomics?

What should the consumer emphasize when seeking involvement in the nutritional genomics arena or in looking at DNA testing for ancestry for practical applications beyond genealogy or archaeogenetics? Look where the money is being spent. More money is spent on advertising for cereals in the United States than is spent by the government on education, according to a video aired July 23, 2003 at television's History Channel titled "American Eats."

Do you eat or remember the typical American diet that includes the old favorites such as Jell-o, Spam, Hershey bars, frozen TV dinners by Swanson and Birdseye, won-ton soup, kielbasa, pizza, pasta, feta cheese, dolmas, and cotton candy? I grew up on those products in the early fifties—in the heart of Brooklyn's Little Italy and Greektown, where the "Mediterranean diet" (Med Diet)—hummos on pita bread competed with spaghetti and tomato sauce and chow mein to go competed with Spam and Twinkies. The old-fashioned typical American diet, influenced by a variety of diverse immigrants gave all of us choices in what foods we ate. Still, we all had food habits and customs that tweaked our core identity or ethnic ancestry in a snack-based world.

Some people still can't view a movie at home or in a theatre without buying snacks and liquid candy-type soda, caffeinated or similar beverages. Some body types use caffeine instead of mild exercise to become more alert. Others use a sugar jolt from a bolt of insulin and adrenalin. And some genotypes are so over-aroused by sensitized nervous systems that one spoonful of sugar or honey or a sip of caffeine causes a panic attack.

A combination of watching horror films and snacking on candy in the movies during the fifties was too much for many delicate nervous systems. Today children are overwhelmed by movie or video special effects that toss their brain and heart rhythms out of sync, sometimes into seizures. Children still drink sugary fruit juices, and the diabetes epidemic is growing, especially among minorities.

When I grew up in the fifties, convenience was king, and we ate the food introduced by all the immigrant groups. In 1953, I'd come home from seventh grade to a frozen Swanson's turkey TV dinner or a burger with mayo, tomato and lettuce on a bun and mashed potatoes followed by a pint of mint chip ice cream almost every day. How about you? There's no way I'm eating like that today. View the excellent videos on American Eats and More American Eats available from http://www.historychannel.com/.

Think about high advertising budgets for snacks advertised a decade ago. What is frequently advertised on TV currently? Drugs—medicines for colds, flu, allergies, high cholesterol, arthritis, and most other chronic illnesses related to years of eating the wrong foods for your genotype. Think of advertising geared to aging boomers. Then there's the shock advertising that jolts you from your alpha state of meditation with loud ambulance sirens and shouts of illness, advertising designed to whip fear into an aging population in order to sell anything from devices to call for emergency help to medicines, through either shock, fear, concern, or worry.

It seems as if most TV commercials today either show people with illness or allergy, or the ads are about drugs for chronic illnesses from high cholesterol to arthritis. Could those chronic illnesses have occurred partly as the result of a one diet-fits-all marketing mentality? You don't see any TV ads yet on eating according to your genotype. Let's see what pops up in the future. Healthcare in advertising is big business. You do see more ads for drugs and exercise equipment, air filters, and emergency electronic response equipment than you do for snacks nowadays, except on the children's shows and cartoons.

Infomercials on exercise equipment run all night along with ads for emergency electronic response communications devices that use shock advertising to scare the wits out of older adults—ambulance sirens, people shouting for help, an opening scene of a white-haired person collapsed on the floor. It certainly frightens older people into changing the station to something less disturbing. These ads come on in the middle of the news, as if the news wasn't stressful enough to watch, but also appear all night long just when older adults who can't sleep are looking for comforting scenes to watch for relaxation.

Think about the old cigarette ads of the fifties on television now replaced by ads for asthma drugs, diabetes blood testing equipment, pet medicines, even vitamins, but vitamins geared for older adults. All those illnesses we see dozens of times a day in the television ads, could they be due to eating the wrong foods or remaining too sedentary? For some genotypes exercise can do more harm than moderate walking a half hour to an hour a day. The food and the exercise would need to be tailored to your body. What about lifestyle changes or stress levels?

Does your body secrete too much stress hormones such as cortisol when you exercise? Changes in diet may not be enough if you're being abused daily by someone, either psychologically, financially, physically, emotionally, or sexually. Maybe you're not allowed to turn on the air conditioner at home or the heat because you're married to someone in good health who's so thrifty that your health isn't being taken into account. Be aware of the variables.

Consider the effects that widespread sugar use had on teeth and on diabetes II among minorities, children, obese persons, and older adults. Is it your food, a virus, or your genes, that contribute to obesity? Are you addicted to the sugar in your cereals, in your "alternative to dairy" beverages? Have you measured the surge of sugar in your bloodstream after drinking a glass of soy, almond, oat, or rice milk with added sweeteners? Is anyone out there drinking Noni juice? I am, by the case, two tablespoons full daily.

Do you make your own unprocessed alternative to dairy products? How do they affect your health? Is your diet too high in soy? Could it be affecting your thyroid? Or is the mercury in the fish you eat causing nervous system problems and hair loss? Have you tried a type of fish less known to have contamination of mercury in amounts unsafe to humans? Do you reach for a can of soda instead of filtered water? Are pesticides affecting your nervous system? How important is convenience for you?

More employers are trimming retiree health care insurance coverage. The cost of providing prescription medicine for older adults is urgent. In light of this, many are turning toward nutritional and genetic research solutions in hope of preventing the need for many different, costly, and sometimes interacting prescription medicines in old age.

Consumers can research studies and information from their Congressional Budget Office. Also see studies such as the one released on July 23, 2003, from the University of Maryland School of Pharmacy that found in the year 2000 only 39 percent of people ages 65 to 69 in the USA had health care insurance from their employers. Coverage is dropping. Is there a nutritional genomic alternative? If your insurance coverage stops upon retirement, can better choices of food play a role in keeping you healthier longer? It's all about choices.

APPENDIX A

▼

NAME FREQUENCY IN THE US

(Reprinted with permission from the U.S. Census Bureau, Population Division)

How Frequently Do Names Appear?

NOTE: No specific individual information is given

See the US Census Bureau Web site at: <u>http://www.census.gov/genealogy/names/</u>

For example:
US Census Bureau, 1990

    (1). A "Name"
    (2). Frequency in percent
    (3). Cumulative Frequency in percent
    (4). Rank

In the file (dist.all.last) one entry appears as:

MOORE   0.312   5.312   9

In our Search Area sample, MOORE ranks 9th in terms of frequency. 5.312 percent of the sample population is covered by MOORE and the 8 names occurring more frequently than MOORE. The surname, MOORE, is possessed by 0.312 percent of our population sample.

Detailed Methodology

Variables in Names Files:
  name
  freq = Frequency in percent
  cum.freq = Cumulative Frequency in percent
  rank

First ten entries in dist.all.last

Name	Freq	cum.freq	Rank
SMITH	1.006	1.006	1
JOHNSON	0.810	1.816	2
WILLIAMS	0.699	2.515	3
JONES	0.621	3.136	4
BROWN	0.621	3.757	5
DAVIS	0.480	4.237	6
MILLER	0.424	4.660	7
WILSON	0.339	5.000	8
MOORE	0.312	5.312	9
TAYLOR	0.311	5.623	10

First ten entries in dist.female.first

Name	Freq	cum.freq	Rank
MARY	2.629	2.629	1
PATRICIA	1.073	3.702	2
LINDA	1.035	4.736	3
BARBARA	0.980	5.716	4
ELIZABETH	0.937	6.653	5
JENNIFER	0.932	7.586	6
MARIA	0.828	8.414	7
SUSAN	0.794	9.209	8
MARGARET	0.768	9.976	9
DOROTHY	0.727	10.703	10

First ten entries in dist.male.first

Name	Freq	cum.freq	Rank
JAMES	3.318	3.318	1
JOHN	3.271	6.589	2
ROBERT	3.143	9.732	3
MICHAEL	2.629	12.361	4
WILLIAM	2.451	14.812	5
DAVID	2.363	17.176	6
RICHARD	1.703	18.878	7
CHARLES	1.523	20.401	8
JOSEPH	1.404	21.805	9
THOMAS	1.380	23.185	10

Source: U.S. Census Bureau, Population Division,
Population Analysis & Evaluation Staff
Maintained By: Laura K. Yax (Population Division)
Last Revised: December 20, 1999 at 11:33:07 AM

APPENDIX B

▼

ETHNIC GENEALOGY WEB SITES:

Acadian/Cajun http://www.acadian.org/tidbits.html
& French Canadian
African-American: http://www.cyndislist.com/african.htm
African Royalty Genealogy: http://www.uq.net.au/~zzhsoszy/
Asia and the Pacific: http://www.cyndislist.com/asia.htm
Austria-Hungary Empire: http://feefhs.org/ah/indexah.html
Chinese Genealogy: http://www.chineseroots.com.
Croatia Genealogy Cross Index http://feefhs.org/cro/indexcro.html
Eastern Europe: http://www.cyndislist.com/easteuro.htm
Eastern European Genealogical Society, Inc.: http://feefhs.org/ca/frg-eegs.html
Ethnic, Religious, and National Index 14 countries: http://feefhs.org/ethnic.html
German Genealogical Digest: http://feefhs.org/pub/frg-ggdp.html
Greek Genealogy Sources on the Internet: http://www-personal.umich.edu/~cgaunt/greece.html
Genealogy Societies Online List: http://www.daddezio.com/catalog/grkndx04.html

Greek Genealogy (Hellenes-Diaspora Greek Genealogy): http://www. geocities.com/SouthBeach/Cove/4537/
Greek Genealogy Home Page: http://www.daddezio.com/grekgen.html
Greek Genealogy Articles: http://www.daddezio.com/catalog/grkndx01.html
India Genealogy: http://genforum.genealogy.com/india/
India Family Histories: http://www.mycinnamontoast.com/perl/ results.cgi?region=79&sort=n
India-Anglo-Indian/Europeans in India genealogy: http://members. ozemail.com.au/~clday/
Irish Travellers: http://www.pitt.edu/~alkst3/Traveller.html
Japanese Genealogy: http://www.rootsweb.com/~jpnwgw/
Jewish Genealogy: http://www.jewishgen.org/infofiles/
Lebanese Genealogy: http://www.rootsweb.com/~lbnwgw/
Melungeon: http://www.geocities.com/Paris/5121/melungeon.htm
Middle East Genealogy: http://www.rootsweb.com/~mdeastgw/index.html
Middle East Genealogy by country: http://www.rootsweb.com/~mdeastgw/ index.html#country
Native American: http://www.cyndislist.com/native.htm
Polish Genealogical Society of America: http://feefhs.org/pol/frg-pgsa.html
Quebec and Francophone: http://www.francogene.com/quebec/amerin.html
Syrian and Lebanese Genealogy: http://www.genealogytoday.com/family/ syrian/
Syria Genealogy: http://www.rootsweb.com/~syrwgw/
Tibetan Genealogy: http://www.distantcousin.com/Links/Ethnic/China/ Tibetan.html
Turkish Genealogy Discussion Group: http://www.turkey.com/forums/ forumdisplay.php3?forumid=18
Unique Peoples: http://www.cyndislist.com/peoples.htm

- **The Unique People's list includes:**

- Black Dutch

- Doukhobors

- Gypsy, Romani, Romany & Travellers

- Melungeons

- Metis

- Miscellaneous

- Wends/Sorbs

Genealogy, (General):

Ancestry.com: http://www.ancestry.com/main.htm?lfl=m
Cyndi's List of Genealogy on the Internet: http://www.cyndislist.com/
Cyndi's List is a categorized & cross-referenced index to genealogical resources on the Internet with thousands of links.
DistantCousin.com (Uniting Cousins Worldwide) http://distantcousin.com/Links/surname.html
Ellis Island Online: http://www.ellisisland.org/
Family History Library: http://www.familysearch.org/Eng/default.asp
http://www.familysearch.org/Eng/Search/frameset_search.asp
(The Church of Jesus Christ of Latter Day Saints) International Genealogical Index
Female Ancestors: http://www.cyndislist.com/female.htm
Genealogist's Index to the Web:
http://www.genealogytoday.com/GIWWW/?
Genealogy Web http://www.genealogyweb.com/
Genealogy Authors and Speakers: http://feefhs.org/frg/frg-a&l.html
Genealogy Today: http://www.genealogytoday.com/
My Genealogy.com http://www.genealogy.com/cgi-bin/my_main.cgi
Scriver, Dr. Charles: The Canadian Medical Hall of Fame http://www.virtualmuseum.ca/Exhibitions/Medicentre/en/scri_print.htm
Surname Sites: http://www.cyndislist.com/surn-gen.htm
National Genealogical Society: http://www.ngsgenealogy.org/index.htm
United States List of Local by State Genealogical Societies: http://www.daddezio.com/society/hill/index.html
United States Vital Records List: http://www.daddezio.com/records/room/index.html or http://www.cyndislist.com/usvital.htm

APPENDIX C

▼

BIBLIOGRAPHY

BIBLIOGRAPHY 1:
Nutritional Genomics:

See the articles on the Web at Nutrition.org at: http://www.nutrition.org/cgi/content/abstract/129/11/1945.

See Research Diets at: http://www.researchdiets.com/.

Human Genome

Adams, Julian, (editor), **Convergent Issues in Genetics and Demography**, (Oxford University Press, New York, 1990).

Brody, Tom, **Nutritional Biochemistry**, (Academic Press, San Diego, 2nd edition, 1999). There is a good discussion about "Nutrition methodology", including dot blots, molecular cloning, and DNA sequencing.

Brooks, William Keith, **The Law of Heredity: A Study of the Cause of Variation, and the Origin of Living Organisms**, (J. Murphy & Company, Baltimore, 1883).

Brown, Michael S., **Time Bombs in the Human Genome: Exploding Triplets That Cause Disease**, Medical Grand Rounds series, (University of Texas Southwestern Center, Dallas, 1993). This book is about the amplification of DNA trinucleotide repeats within the human genome. This can cause diseases such as Myotonic Dystrophy, Fragile X, and Muscular Atrophy.

Buckley, John J., **Genetics Now: Ethical Issues in Genetic Research**, (University Press of America, Washington, D.C., 1978).

Zallen, Doris Teichler, **Does It Run in the Family? A Consumer's Guide to DNATesting for Genetic Disorders**, (Rutgers University Press, New Brunswick, New Jersey, 1997).

Zlinskas, Raymond A, and Balint, Peter J., (editors), **The Human Genome-Project and Minority Communities: Ethical, Social, and Political Dilemmas**, (Praeger, Westport, Conneticut, 2000).

Classics in Molecular Genetics:

Watson, James D. **Molecular Biology of the Gene**, (W.A. Benjamin, Inc., New York, 1965, 2nd edition 1971, 3rd edition 1978, 4th edition 1987). This is the first edition of the "classic" textbook on molecular biology. At the time it was written, the genetic code was just being completely deciphered.

Watson, James D. **The Double Helix**, (W.A. Benjamin, Inc., New York, 1968, the "Norton Critical Edition" of this was published in 1980, edited by Gunther Stent, with reprints of some of the original papers.). This is a fun, very readable, "gossipy" account of the discovery of the DNA double helix. Of course, it is very biased, but nonetheless it makes good reading, and is very insightful into the personalities and politics involved in solving the basic structure of DNA.

Watson, J.D. and Tooze, J., **THE DNA STORY—A Documentary History of Gene Cloning** (W.H. Freeman and Company, San Francisco, 1981). This is kind of like a "scrapbook", with newspaper clippings, old letters, and lots of interesting stories about the beginning of genetic engineering.

The Human Genome in Europe: Scientific, Ethical, and Social Aspects, (Academie Royale de Medicine de Belgique, Bruxelles, 1995). This book is based on the proceedings of a conference in Brussels, Palais des academies, 18-19 September, 1995, by the Commission of the European Communities. Also see a nutrigenomics bibliography Web site at: http://food.oregonstate.edu/nutri.html.

*   *   *   *

BIBLIOGRAPHY 2 : GENERAL:
DNA Testing and Genetics.

A Biologist's Guide to Analysis of DNA Microarray Data Steen Knudsen/ Hardcover/Wiley, John & Sons, Incorporated/April 2002

Advances and Opportunities in DNA Testing and Gene Probes Business Communications Company Incorporated (Editor)/Hardcover/Business Communications/September 1996

African Exodus, The Origins of Modern Humanity Stringer, Christopher and Robin McKie. Henry Holt And Company 1997

An A to Z of DNA Science: What Scientists Mean when They Talk about Genes and Genomes Jeffre L. Witherly, Galen P. Perry, Darryl L. Leja/Paperback/Cold Spring Harbor Laboratory Press/September 2002

An Introduction to Forensic DNA Analysis Norah Rudin, Keith Inman/Hardcover/CRC Press/December 2001

Archaeogenetics: DNA and the population prehistory of Europe, Ed. Colin Renfrew & Katie Boyle. McDonald Institute Monographs. Cambridge, UK, Distributed by Oxbow Books UK. In USA: The David Brown Book Company, Oakville, CT. 2000

Cartoon Guide to Genetics Gonick, Larry, With Mark Wheelis: Paperback/ HarperInformation/July 1991

DNA Detectives, The—Working Against Time, novel, Hart, Anne. Mystery and Suspense Press, iuniverse.com paperback 248 pages at http://www. iuniverse.com or 1-877-823-9235.

DNA for Family Historians (ISBN 0-9539171-0-X). Savin, Alan of Maidenhead, England, is author of the 32-page book. See the Web site: http:// www.savin.org/dna/dna-book.html

DNA Microarrays and Gene Expression Pierre Baldi, G. Wesley Hatfield, G. Wesley Hatfield/Hardcover/Cambridge University Press/August 2002

Microarrays for an Integrative Genomics Isaac S. Kohane, Alvin Kho, Atul J. Butte/Hardcover/MIT Press/August 2002

Does It Run in the Family?: A Consumers Guide to DNA Testing for Genetic Disorders Doris Teichler Zallen, Doris Teichler-Zallen, Doris Teichler Zallen/Hardcover/Rutgers University Press/May 1997

Double Helix, The: A Personal Account of the Discovery of the Structure of DNA James D. Watson/Paperback/Simon & Schuster Trade Paperbacks/June 2001

Genes, Peoples, and Languages Luigi Luca Cavalli-Sforza, Mark Seielstad (Translator).

Genetic Witness: Forensic Uses of DNA Tests DIANE Publishing Company (Editor)/Paperback/DIANE Publishing Company/April 1993

History and Geography of Human Genes, The [ABRIDGED] L. Luca Cavalli-Sforza, Paolo Menozzi (Contributor), Alberto Piazza (Contributor).

How to DNA Test Our Family Relationships Terry Carmichael, Alexander Ivanof Kuklin, Ed Grotjan/Paperback/Acen Press/November 2000

Introduction to Genetic Analysis Anthony J. Griffiths, Suzuki, Lewontin, Gelbart, David T. Suzuki, Richard C. Lewontin, Willi Gelbart, Miller, Jeffrey H. Miller/Hardcover/W. H. Freeman Company/February 2000

Jefferson's Children: The Story of One American Family Shannon Lanier, Jane Feldman, Lucian K. Truscott (Introduction)/Hardcover/Random House Books for Young Readers/September 2000

Medical Genetics Lynn B. B. Jorde, Michael J. Bamshad, Raymond L. White, Michael J. Bamshad, John C. Carey, John C. Carey, Raymond L. White, John C. Carey/Paperback/Mosby-Year Book, Inc./July 2000

Molecule Hunt, The: Archaeology and the Search for Ancient DNA Martin Jones/Hardcover/Arcade/April 2002

More Chemistry and Crime: From Marsh Arsenic Test to DNA Profile Richard Saferstein, Samuel M. Gerber (Editor)/Hardcover/American Chemical Society/August 1998

1996, Quest For Perfection—The Drive to Breed Better Human Beings, Maranto, Gina. Scribner, 1996

Our Molecular Future: How Nanotechnology, Robotics, Genetics, and Artificial Intelligence Will Transform Our World Mulhall, Douglas./Hardcover/ Prometheus Books/March 2002

Paternity—Disputed, Typing, PCR and DNA Tests: Index of New Information Dexter Z. Franklin/Hardcover/Abbe Pub Assn of Washington Dc/January 1998

Paternity in Primates: Tests and Theories R. D. Martin (Editor), A. F. Dickson (Editor), E. J. Wickings (Editor)/Hardcover/Karger, S Publishers/December 1991

Queen Victoria's Gene: Hemophilia and the Royal Family (Pbk) D. M. Potts, W. T. Potts/Paperback/Sutton Publishing, Limited/June 1999

Redesigning Humans: Our Inevitable Genetic Future Stock, Gregory./Hardcover/Houghton Mifflin Company/April 2002

Rosalind Franklin: The Dark Lady of DNA, Brenda Maddox/Hardcover/HarperCollins Publishers/October 2002

Schaum's Outline Of Genetics Susan Elrod, William D. Stansfield/Paperback/ McGraw-Hill Companies, The/December 2001

Seven Daughters of Eve, The: The Science That Reveals Our Genetic Ancestry. Sykes, Bryan. **ISBN:** 0393323145 **Publisher**: Norton, W. W. & Company, Inc. May 2002

Stedman's OB-GYN & Genetics Words Ellen Atwood (Editor), Stedmans/ Paperback/Lippincott Williams & Wilkins/December 2000

*   *   *   *

BIBLIOGRAPHY 3:
Genealogy.

A Bintel Brief: Sixty Years of Letters From the Lower East Side to the Jewish Daily Forward. Metzker, Isaac, ed Doubleday and Co. 1971. Garden City, NY

Climbing Your Family Tree: Online and Offline Genealogy for Kids IRA Wolfman, Tim Robinson (Illustrator), Alex Haley (Introduction)/Paperback/ Workman Publishing Company, Inc./October 2001

Complete Beginner's Guide to Genealogy, the Internet, and Your Genealogy Computer Program Karen Clifford/Paperback/Genealogical Publishing Company, Incorporated/February 2001

Complete Idiot's Guide(R) to Online Geneology Rhonda McClure/Paperback/Pearson Education/January 2002

Creating Your Family Heritage Scrapbook : From Ancestors to Grandchildren, Your Complete Resource & Idea Book for Creating a Treasured Heirloom. Nerius, Maria Given, Bill Gardner ISBN: 0761530142 Published by Prima Publishing, Aug 2001

Cyndi's List: A Comprehensive List of 70,000 Genealogy Sites on the Internet (Vol. 1 & 2) Cyndi Howells/Paperback/Genealogical Publishing Company, Incorporated/June 2001.

Discovering Your Female Ancestors: Special strategies for uncovering your hard-to-find information about your female lineage. Carmack, Sharon DeBartolo. Conference Lecture on Audio Tape: Carmack, Sharon DeBartolo.

Folklife and Fieldwork: A Layman's Introduction to Field Techniques. Bartis, Peter. Washington, DC: Library of Congress, 1990.

Genealogy Online for Dummies Matthew L. Helm, April Leigh Helm, April Leigh Helm, Matthew L. Helm/Paperback/Wiley, John & Sons, Incorporated/ February 2001

Genealogy Online Elizabeth Powell Crowe/Paperback/McGraw-Hill Companies, November 2001

History From Below: How to Uncover and Tell the Story of Your Community, Association, or Union. Brecher, Jeremy. New Haven: Advocate Press/ Commonwork Pamphlets, 1988.

My Family Tree Workbook: Genealogy for Beginners Rosemary A. Chorzempa/Paperback/Dover Publications, Incorporated/

National Genealogical Society Quarterly 79, no. 3 (September 19991): 183-93

"Numbering Your Genealogy: Sound and Simple Systems." Curran, Joan Ferris.

Oral History and the Law. Neuenschwander, John. Pamphlet Series #1. Albuquerque: Oral History Association, 1993.

Oral History for the Local Historical Society. Baum, Willa K. Nashville: American Association for State and Local History, 1987.

Scrapbook Storytelling: Save Family Stories & Memories with Photos, Journaling & Your Own Creativity Slan, Joanna Campbell, Published by EFG, Incorporated, ISBN: 0963022288 May 1999

"The Silent Woman: Bringing a Name to Life." NE-59. Boston, MA: New England Historic Genealogical Society Sesquicentennial Conference, 1995.

The Source: A Guidebook of American Genealogy Alice Eichholz, Loretto Dennis Szucs (Editor), Sandra Hargreaves Luebking (Editor), Sandra Hargreaves Luebking (Editor)/Hardcover/MyFamily.com, Incorporated/February 1997

To Our Children's Children: Journal of Family Members, Bob Greene, D. G. Fulford 240pp. ISBN: 038549064X Publisher: Doubleday & Company, Incorporated: October 1998.

Transcribing and Editing Oral History. Nashville: American Association for State and Local History, 1991.

Using Oral History in Community History Projects. Buckendorf, Madeline, and Laurie Mercier. Pamphlet Series #4. Albuqueque: Oral History Association, 1992.

Unpuzzling Your Past: The Best-Selling Basic Guide to Genealogy (Expanded, Updated and Revised) Emily Anne Croom, Emily Croom/Paperback/F & W Publications, Incorporated/August 2001

Writing a Woman's Life. Heilbrun, Carolyn G. New York: W.W. Norton, 1988

Your Guide to the Family History Library: How to Access the World's Largest Genealogy Resource Paula Stuart Warren, James W. Warren/Paperback/F & W Publications, Incorporated/August 2001

Your Story: A Guided Interview Through Your Personal and Family History, 2nd ed., 64pp.ISBN: 0966604105 Publisher: Stack Resources, LLC

*   *   *   *

BOOKS WRITTEN BY ANNE HART

1. **How to Interpret Your DNATest Results for Family History & Ancestry**
   ISBN: 0-595-26334-8—Paperback—List Price:$19.95
   Publisher: Writers Club Press, iUniverse, Incorporated. Published Date: 12/20/2002. CD Audio available directly from author. To order paperback version, click on iUniverse Bookstore at www.iuniverse.com. Or phone: **1-877-823-9235**

2. **The DNA Detectives—Working Against Time** (romantic suspense novel)
   ISBN: 0-595-25339-3—Paperback—List Price: $16.95
   Publisher: iUniverse, Incorporated—Published Date: 10/11/2002

3. **Anne Joan Levine, Private Eye**
   ISBN: 0595218601—Paperback—List Price: $16.95
   Publisher: iUniverse, Incorporated—Published Date: 03/01/2002
   Author: Anne Hart

4. **Astronauts and Their Cats**
    ISBN: 0595223303—Paperback—List Price: $11.95
    Publisher: iUniverse, Incorporated—Published Date: 04/01/2002
    Author: Anne Hart

5. **Cleopatra's Daughter**
    ISBN: 0595220215—Paperback—List Price: $11.95
    Publisher: iUniverse, Incorporated—Published Date: 03/01/2002
    Author: Anne Hart

6. **Counseling Anarchists**
    ISBN: 0595220541—Paperback—List Price: $14.95
    Publisher: iUniverse, Incorporated—Published Date: 03/01/2002
    Author: Anne Hart

7. **Courage to be Jewish & the Wife of an Arab Sheik**
    ISBN: 0595187900—Paperback—List Price: $21.95
    Publisher: iUniverse, Incorporated—Published Date: 06/01/2001
    Author: Anne Hart

8. **Cyber Snoop Nation**
    ISBN: 0595220339—Paperback—List Price: $13.95
    Publisher: iUniverse, Incorporated—Published Date: 03/01/2002
    Author: Anne Hart

9. Cyberscribes 1:The New Journalists (out of print)
    ISBN: 1880663651—Paperback—List Price: $24.95
    Publisher: Ellipsys International Publications, Incorporated—Published
    Date: 05/01/1997
    Author: Anne Hart

10. **Date Who Unleashed Hell**
    ISBN: 0595219829—Paperback—List Price: $17.95
    Publisher: iUniverse, Incorporated—Published Date: 03/01/2002
    Author: Anne Hart

11. **Day My Whole Country Turned Jewish**
    ISBN: 0759663807—Paperback—List Price: $18.95
    Publisher: 1stBooks Library—Published Date: 06/01/2002
    www.1stbooks.com
    Author: Anne Hart

12. **Four Astronauts and a Kitten**
    ISBN: 0595192025—Paperback—List Price: $11.95
    Publisher: iUniverse, Incorporated—Published Date: 07/01/2001
    Author: Anne Hart

13. **Freelance Writer's E-Publishing Guidebook**
    ISBN: 0595189520—Paperback—List Price: $36.95
    Publisher: iUniverse, Incorporated—Published Date: 06/01/2001
    Author: Anne Hart

14. **How to Make Money Organizing Information**
    ISBN: 0595236952—Paperback—List Price: $35.95
    Publisher: iUniverse, Incorporated—Published Date: 08/01/2002
    Author: Anne Hart

15. **How to Make Money Teaching Online with Your Camcorder and PC**
    ISBN: 0595221238—Paperback—List Price: $36.95
    Publisher: iUniverse, Incorporated—Published Date: 03/01/2002
    Author: Anne Hart

16. **How to Stop Elderly Abuse**
    ISBN: 0595235506—Paperback—List Price: $28.95
    Publisher: iUniverse, Incorporated—Published Date: 07/01/2002
    Author: Anne Hart

17. **How Two Yellow Labs Saved the Space Program**
    ISBN: 0595231810—Paperback—List Price: $11.95
    Publisher: iUniverse, Incorporated—Published Date: 06/01/2002
    Author: Anne Hart

18. **Khazars Will Rise Again!**
    ISBN: 059521830X—Paperback—List Price: $15.95
    Publisher: iUniverse, Incorporated—Published Date: 03/01/2002
    Author: Anne Hart

19. <u>**How to Make Money Selling Facts to Nontraditional Markets**</u>
    ISBN: 0-595-27842-6—Paperback—List Price: $33.95
    Publisher: iUniverse, Incorporated—Published Date: 03/17/2003
    Author: <u>Anne Hart</u>

20. **Make Money with Your Camcorder and PC**
    ISBN: 0595218644—Paperback—List Price: $36.95
    Publisher: iUniverse, Incorporated—Published Date: 04/01/2001
    Author: <u>Anne Hart</u>

21. **Murder in the Women's Studies Department**
    ISBN: 0595218598—Paperback—List Price: $16.95
    Publisher: iUniverse, Incorporated—Published Date: 03/01/2002
    Author: <u>Anne Hart</u>

22. **New Afghanistan's TV Anchorwoman**
    ISBN: 0595215572—Paperback—List Price: $23.95
    Publisher: iUniverse, Incorporated—Published Date: 02/01/2002
    Author: <u>Anne Hart</u>

23. **Power Dating Games**
    ISBN: 059519186X—Paperback—List Price: $23.95
    Publisher: iUniverse, Incorporated—Published Date: 07/01/2001
    Author: <u>Anne Hart</u>

24. **Private Eye Called Mama Africa**
    ISBN: 0595189407—Paperback—List Price: $25.95
    Publisher: iUniverse, Incorporated—Published Date: 06/01/2001
    Author: <u>Anne Hart</u>

25. <u>**Roman Justice: SPQR—Too Roman to Handle**</u>
    ISBN: 0-595-27282-7-Paperback—List Price: $13.95
    Publisher: iUniverse, Incorporated—Published Date: 03/17/2003
    Author: <u>Anne Hart</u>

26. **Sacramento Latina**
    ISBN: 0595220614—Paperback—List Price: $29.95
    Publisher: iUniverse, Incorporated—Published Date: 03/01/2002
    Author: <u>Anne Hart</u>

27. **Tools for Mystery Writers**
    ISBN: 0595217478—Paperback—List Price: $28.95
    Publisher: iUniverse, Incorporated—Published Date: 03/01/2002
    Author: Anne Hart

28. **Verbal Intercourse**
    ISBN: 0595219462—Paperback—List Price: $17.95
    Publisher: iUniverse, Incorporated—Published Date: 03/01/2002
    Author: Anne Hart

29. Winning Resumes for Computer Personnel
    ISBN: 0764101307—Paperback—List Price: $12.95
    Publisher: Barrons Educational Series, Incorporated—Published Date: 02/
    01/1998
    Author: Anne Hart

30. **Writer's Bible**
    ISBN: 0595193056—Paperback—List Price: $28.95
    Publisher: iUniverse, Incorporated—Published Date: 08/01/2001
    Author: Anne Hart

31. **Writing What People Buy**
    ISBN: 0595219365—Paperback—List Price: $24.95
    Publisher: iUniverse, Incorporated—Published Date: 03/01/2002
    Author: Anne Hart

32. **The Year My Whole Country Turned Jewish**
    ISBN: 0759672512—Paperback—List Price: $14.95
    Publisher: 1stBooks Library—Published Date: 03/01/2002
    www.1stbooks.com
    Author: Anne Hart

33. **Winning Resumes for Computer Personnel. 2 nd Ed.**
    Barron's Educational Series, 1994, 1998. (Out of print March 2003.)
    Author: Anne Hart

34. In the Chips: **How to Make Money with Your PC**, 1985, Simon &
    Schuster (out of print).

35. **High Paying Jobs in Six Months or Less**, 1984, Simon & Schuster, Monarch Books. (out of print).

36. **Careers in Robotics**, Simon & Schuster, Arco Books, 1985. (out of print)

37. **Careers in Aerospace**, Simon & Schuster, Arco Books, 1985. (out of print)

38. **Robotics**, Tab Books, 1985 (out of print)

39. **Opportunities in Home Health Care Careers**, VGM Books, 1992.

40. **Winning Tactics for Women Over 40** (financial education) 1988, Mills & Sanderson, primary author on co-authored book. (out of print)

41. **Cyberlife**, author 57-page chapter on ***Virtual Reality Careers***, of multi-authored book. Sams, 1995.

42. **Cyberscribes.1 The New Journalists**, Ellipsys International Publishers, 1996 (out of print) (Book selected as college textbook in graduate schools of new media studies and chosen as #15 for week in May on Ingram's computer book best sellers list.)

43. Wrote 23 published bioscience, how-to, social science, health, and career development pamphlets on various issues in the news, each book between 50 and 100 pages for Pamphlet Publications, Inc., from 1978 to 1994.

44. **Tracing Your Jewish DNA for Family History & Ancestry: Merging a Mosaic of Communities.** Writer's Club Press, iUniverse.com at http:// www.iuniverse.com. Paperback. 2003.

45. **The Beginner's Guide to Interpreting Ethnic DNA Origins for Family History: How Ashkenazi, Sephardi, Mizrahi & Europeans Are Related to Everyone Else.** ISBN: 0-595-28306-3. Paperback. Writer's Club Press, iUniverse.com at http://www.iuniverse.com. Published: July 7-2003.

46. **Nutritional Genomics—A Consumer's Guide to How Your Genes and Ancestry Respond to Food.** Paperback. Writer's Club Press, iUniverse.com at http://www.iuniverse.com. Published Aug. 2003.

APPENDIX D

▼

WHERE TO FIND SPEECH TRANSCRIPTS

Consumers can dialogue with the food industry. Brainstorm how you can work together to research and communicate about the "shift in thinking about the way food is processed, packaged, and marketed in the U.S," according to an online report in the Harvard Public Health NOW, published biweekly. Meetings and a July 2003 conference drew the food industry, public health professionals, scientists, nutrition researchers, and anyone else present. FDA has critical issues to handle—such as safety and security of food and edible imports—in addition to helping consumers improve nutrition. That's why FDA added more than 650 new food safety monitors in the field.

In a speech before Harvard School of Public Health, Remarks by Mark B McClellan, MD, PhD, Commissioner, Food and Drug Administration presented on July 1, 2003 included, "We need your input on how we can use these new means to achieve the greatest improvements in public health protection, with the least possible disruption and added costs for the foods we regulate. And we must do it quickly, because these new measures have to be implemented before the end of the year."

Consumers—your help is needed. Here's how to find out more information. Research speech archives and obtain a transcript of the Speech before Harvard School of Public Health, Remarks by Mark B McClellan, MD, PhD, Commissioner, Food and Drug Administration presented on July 1, 2003. For more information about news or conferences, see: Harvard Public Health NOW, published biweekly by the Office of Communications, Harvard School of Public Health, 665 Huntington Ave., SPH 1-1312A, Boston, Massachusetts 02115. Or click on the Web site at: http://www.hsph.harvard.edu/now/jul11/conference.html Search the archives of the Harvard School of Public Health at: http://www.hsph.harvard.edu/now/archives.html.

Contact the Office of Communications, Harvard School of Public Health if you are writing on public health issues for the media. Since numbers could change, click on the Web site at: http://www.hsph.harvard.edu/communications.html to find current names of people to contact, phone, fax, or email addresses that will help you get the expert information you need to write your article.

Contact Information
Director of Communications
Kresge 1015
677 Huntington Avenue
Boston, MA 02115

Media inquiries
Assistant Director of Communications
Kresge 1014
677 Huntington Avenue
Boston, MA 02115

General Mailing Address:
Harvard School of Public Health
Office of Communications
677 Huntington Avenue
Boston, MA 02115

APPENDIX E

▼

ABOUT NUTRITIONAL GENOMICS (ALSO ABBREVIATED IN THE INDUSTRY AS NUTRIGENOMICS).

According to Dr. Fredric D. Abramson, Ph.D, S.M., Esq. President and CEO of AlphaGenics, Inc., in his article titled, "About NutriGenomics", Copyright 2002, 2003 Reprinted here with permission.

About NutriGenomics

Fredric Abramson, Ph.D., S.M., Esq.
Copyright 2002, 2003

Our genes determine who we are, how we develop and age, and how we respond to our environment. They are the blueprints that define our potential. But they do not act in a vacuum. They respond to and actually help shape our environment. That environment includes what we put into our body in food, water, supplements, nutraceuticals and pharmaceuticals.

Together, our genes and environment control our health and our susceptibility to disease. NutriGenomics involves decoding how the molecular composition of our diet, which includes food, supplements, pharmaceuticals and water) influences the work being done by our genes, and then defining personalized dietary strategies to tune each person's gene expression pattern, called the "Expressitype." This tuning process must be changed dynamically as the person's genetic activity changes.

"NutriGenomics lets people pick the foods they eat based on how well the foods make their genes function."

Because what we eat influences our health and the way our genes work, we have the opportunity to let people control their health destiny in a whole new way. This is what NutriGenomics is about: calibrating the mix and amounts of what ingredients we put into our body so that our genes work at their best, for our best health status. By building on each person's genetic uniqueness, NutriGenomics focuses on what that individual's genes are doing and how that person can pick and choose things from their environment that will make his/her genes work better or worse.

AlphaGenics approach is to identify what parts of your diet are making your genes work badly, and determine the best mixture in your diet to calibrate and tune the work your genes are doing. Our goal is to adjust your genetic activity, without changing your genes, by dynamically adjusting the mix of what you put into your body as your body changes over time.

Basic Terms

- **Genotype**: The entire set of genes of an individual.
- **Expressitype**: The profile of gene expression activity in an individual at a moment in time.
- **Phenotype**: The observable characteristics of an individual.
- **NutriGenomics**: The science that relates how the molecular inputs from a person's environment influence and control gene expression.

> • **Qwink**: A molecule or particle in the environment that can change gene expression directly or indirectly.
> • **Expressitype Knowledge Transfer System**: A proprietary information/ knowledge management system that integrates individual longitudinal data with a variety of scientifically verified data elements. Its outputs include personalized, dynamic dietary strategies as well as a variety of scientific and research-oriented solutions.

Our dietary environment influences how our genes work

Many different ingredients exist in a person's dietary environment. They exist as both natural molecules and artificial compounds and are found in food as well as in our water, and in dietary supplements, cosmetics and medications. Even in the air we breathe. Some are present by design; others as a by-product. So, for example, the common tomato is known to have over 300 different natural molecules. Some ingredients are man-made, but most are natural substances whose variety and health-effects are not even fully explored.

The key to NutriGenomics is identifying how specific compounds or chemicals in our diet influence the activity of one or more genes, and hence our health.

Our focus is on our genetic activity (Expressitype) and not just what genes we have (Genotype). Knowing your genotype is a lot like knowing where the entrance is to an interstate highway. It is very useful and important, but it doesn't tell you what direction to go in or when you will be able to stop for gas.

The term "Qwink" was coined to refer to any molecule in the dietary stream that influences gene expression. Qwinks are found in foods, supplements, cosmetics, water and pharmaceutical compounds. They include basic ingredients such as sugars, proteins and fat as well as very specialized substances such as vitamins or toxins. A Qwink may be a natural substance or synthetic.

A Qwink is neutral as to whether it has a positive or negative effect. If the gene expression moves in the direction to provide better health or reduce risk, it is good; if it moves to worsen a person's health status, it is bad. Thus, it is only the effects that are good or bad. For example, a pesticide residue that survives in the food chain and influences gene expression is a Qwink.

While pesticides are generally considered harmful, interestingly enough, there is some evidence that certain pesticides could actually reduce the risk of cancer, at least in model systems.[1]

NutriGenomics calibrates dietary inputs and patterns to provide different
Qwink types and concentrations to the genetic machinery.

The impact of a Qwink typically depends on its concentration and dosage frequency.

One of the amazing aspects of human biology is that it responds differently to
varied Qwink concentrations. Take arsenic, for example. Most people know that
arsenic in high doses can kill. What is less commonly known, however, is that at
very low concentrations, arsenic is considered a nutrient.

The same will be found for how Qwinks work. At very low concentrations,
one set of genes may be affected while a much higher concentration may change
expression in an entirely different set of genes.

NutriGenomics is personalized genomics

The way a person's dietary environment influences gene expression and health is
called NutriGenomics. NutriGenomics is a scientific platform that permits
focused, personalized adjustments to a person's dietary environment. An advantage of NutriGenomics is that it is a unifying standard that includes the effect of
dietary supplements, nutraceuticals, pharmaceuticals and cosmetics in addition to
what a person normally eats, as well as environmental factors such as toxins, con-
taminants and infectious agents, such as viruses and bacteria. The scientific goal is
to identify molecular levers coming from a person's environment that can move
the person's gene expression patterns in an appropriate direction, and to further
identify environmental adjustments that can help improve the person's health
status through changes in gene expression.

Our core scientific proposition is that evaluating and modulating a person's
Expressitype as it dynamically changes over time is a more potent and acceptable
way to prevent and control disease than working with genotype alone. The practical difference between focusing on genotype versus expressitype is comparable
to the difference between buying a car based on what Consumer Reports says is
its repair frequency and listening to your car coughing and sputtering when you
are driving down the highway.

---

1. John Milner, National Cancer Institute, personal communication, January
   2003.

The cycle begins with a person's environment, which interacts with the person's genotype and can up-regulate or down-regulate various genes. The gene expression pattern, the expressitype, in turn, works to produce proteins and otherwise control the person's metabolism and physiology to produce the phenotype. It is a person's phenotype at which we observe health and illness. Examples of phenotype include eye color, blood groups, and various chronic and acute diseases. A person's phenotype in turn helps shape the environment. For example, phenotype can influence what a person chooses to eat. A person who doesn't feel well may eat differently; a person who changes their physical activity may also change how they eat; and so on.

AlphaGenics designed its research and intervention around this cycle. By capturing data and information from each individual, including serial measures of expressitype, and blending this data with scientific knowledge from genomics, nutrition, pharmacology, toxicology, and medicine, we can focus on unraveling the systems biology map of how the complex environment interacts with and influences the comparably complex genome.

Our genome works as a system that takes us through our life cycle

Our genetic apparatus is a complex system designed to sustain our lives in intimate communication with the environment, from birth through puberty and the gradual aging process leading to the end of our lives. It is this complexity that has made it so difficult to find cures or treatments for so many diseases like cancer and heart disease. It is this same complexity that explains why a drug will work in one person, be ineffective in another, and harm a third. It is this complexity that confirms why different people need entirely different diets to lose weight, for example.

The fact is also that almost all of the major health conditions that concerns society today–cancer, heart disease, diabetes, obesity, neurologic disorders and so on are all multifactorial and polygenic. Multifactorial means that it is both environment and genetics in combination that explains when, why and where different diseases occur. Polygenic means that multiple genes are involved, not just one.

Three specific aspects of this complexity are worth exploring. First, the number of different genes involved with each of the diseases that plague us can be in the hundreds or even thousands. Second, virtually every chemical that we put into our body influences some genes. And third, our genes typically work collab-

oratively, where one influences another. Our genes work as an integrated, cohe-
sive system.

A change in gene expression is not always a benefit

The goal of NutriGenomics is to move gene expression patterns to some "best"
state. However, because the human genome is complex, an important issue is that
an "increase or a decrease" in gene expression is not "always" linked with a benefit. A lot may depend on what the other genes involved are doing as well.

Many different genes are involved in the common health conditions

Most of the common conditions that concern our health are polygenic, i.e.
involve many different genes as mentioned above. The goal of the Human
Genome Project was to identify thousands of genes and to link these genes to specific health effects. Scientists are also pinpointing exactly where each of these
genes are located, or "mapped," on our chromosomes.[2]

Some examples include breast cancer (over 200 genes, 39 of which are
mapped to 17 different chromosomes), obesity (over 200 genes; 15 mapped to 11
chromosomes); Diabetes (60 mapped genes to 17 different chromosomes); Lym-
phoma (200 mapped genes on 22 chromosomes); and hypertension (19 genes on
10 chromosomes).[3]

Qwinks impact multiple genes, not just one

What we put into our body typically influences more than one gene. For example, the natural compound retinoic acid can change expression in more than 500
different genes[4] while the enzyme Cofactor Q10 will up-or down-regulate over
100 different genes.[5]

---

2. Humans have 22 pairs of chromosomes plus the X and Y.
3. Obtained from the Human Genome Project web site..
4. Balmer and Blomhoff, Gene Expression Regulation by Retinoic Acid,
   Journal of Lipid Research, Vol. 43, 1773-1808 (November 2002).
5. Linnane AW, et al. Cellular redox activity of CoQ10:effect of Q10 supple.
   On human skel. muscle. Free Rad Res 2002;36(4):445-453.

Genes work together as a coherent system, not as independent actions.

For a polygenic condition, it is important to realize that the genes involved do not work by themselves but work in a coherent, structured system. We can take as an example the more than 200 different genes implicated in obesity. If each of these genes worked independently of one another, there would be 2 to the 200th power different combinations. This is 1.6 times ten to the 60th power; that is, 1.6 followed by 60 zeros. This number, 1.6 times ten to the 60th, is very large. In fact, it is larger than all the humans who have ever lived on the earth. It is larger than all the organisms, including viruses, that have ever lived on the earth.

So clearly, just by observing that different groups or types exist in obesity, one realizes that these 200 genes don't work independently. In fact, what must happen is that when one gene turns on, it in turn regulates ten or twenty others. This interaction among our genes leads to a relatively small, countable number of combinations of gene activity. Instead of the astronomic number calculated above, we have only 100's or 1000's of combinations for this example of obesity. These combinations are the "typical" patterns of gene activation, so to speak the molecular bar codes, each of which could be associated with a form of the disease.

So it is possible to imagine two type of obese people, one with five obesity genes turned on, say numbers 1, 50, 95, 150 and 200, and the rest turned off; the second with twelve different obesity genes turned on. These two different combinations could translate into different obesity effects, where one type of person gains weight easier than the other on the same diet.

While our genes are fixed at birth, the work they do changes as we age

Each person is born with genes they inherited from their parents. The identity of your genes is called the "Genotype." The Genotype is a person's unique collection of genes. Classic genetics described genotypes for hair and eye color, and blood types. For all practical purposes, the genotype is fixed when your parents' sperm and egg unite.

Genes are dynamic engines. Their work varies depending on circumstances and time. The entire genomic system changes its work, as illustrated by the aging process that begins with infancy, moves through childhood, puberty and adulthood, and then to senescence. These genetic mechanisms are starting to be understood.

An example of gradual, almost invisible changes in genetic activity is observed in how menstrual cycles shorten as a woman ages. Research done at the University of Minnesota in the 40's shows that a 40 year old woman has one more menstrual period a year than a 20 year old, and that the one-day per year change in cycle length is linear from 20 to 40. This gradual change suggests that key genetic components to target in NutriGenomics are genes whose activity undergoes very gradual changes over a long time period. This further suggests that many key NutriGenomic interventions will be based on cumulative effects.

It's Gene Expression (Expressitype) Which is Key

The accomplishments of the Human Genome Project mean we will soon be able to identify every gene in a person, in effect, that person's genetic blueprint. Some of these genes will indicate greater risks for certain diseases; other genes will mean less risk than average. An example is the BRCA1 gene for breast cancer. Women with BRCA1 have a significantly higher risk of developing breast cancer than those without it.

The genotype, however, is only a starting place to understand what our genetic destiny might be. This is because the work a gene does is likely to change over time, both according to preprogrammed rules, and in response to environmental influences, notably our food. Some genes are turned on early in life, say during pregnancy, and are switched off permanently. Other genes may just sit there waiting to be activated or to become phenotypically manifest. This appears to be the case with certain diseases that arise later in life, such as Huntington's chorea, a single dominant gene disorder that shows itself in a person's 50's.

Further complicating the situation is that when a gene is "on" it may be operating at one activity level one day and a higher or lower level the next day. A gene's activity is exhibited as "genetic expression," a term that refers to the gene producing RNA and otherwise interacting with a person's metabolism and physiology.

Let's call, in analogy to Genotype, the entire profile of expression of each gene throughout the genome, the "Expressitype." Thus, for any health condition, each person has his or her own Expressitype. The Expressitype is the actual amount of genetic activity produced by each gene that is implicated in or related to a specific health status or condition.

Since for any particular gene that is turned on, the expression level may be changing over time, and because the gene itself may be turned on or off at a particular point in time, a person's Expressitypes is likely to change.

A person's NutriGenomic profile, then, measures what each gene is doing at a point in time. By stringing together a series of NutriGenomic profiles, it is possible to see whether a gene is changing, the direction a gene is going and how fast it is getting there.

The gene-environment interaction is very powerful

The interaction between genes and environment is undisputed. Part of the environment is dietary—what we eat. Another part is lifestyle—how we live our everyday lives. Figure 1 is a general overview of NutriGenomics, starting with the dietary inputs at the base and moving through the genome to the end result of health or illness.

Even well defined genetic disorders respond to environment differences. For example, if identical twins each have the single dominant gene for Huntington's chorea, and the twins have different dietary and exercise programs, the onset of Huntington's chorea can be delayed up to eight years in one of the twins.

Diet has been shown to alter gene expression in several ways Qwinks can act on how DNA is transcribed into RNA They can be involved in various metabolic pathways and can increase or decrease the concentrations of other materials needed by the genome. In other words, Qwink effects can be direct, by changing gene activity itself, or indirect by shifting what is available for the gene to use.

Foods, supplements and other chemical sources are the foundation of NutriGenomics. The ingredients that actively regulate the genome, the Qwinks, can turn genes on and off, and can change the amount of activity for each turned-on gene. The Qwinks help determine the Expressitype, which is the profile of what each gene is doing at a moment in time. Gene expression becomes translated into health, wellness or illness through a person's metabolism and the many proteins in the proteome. Moreover, a person's health status, in turn, influences gene activity and can even change what might be acceptable to eat. An example would be creating an immune response to a certain food (i.e., strawberries) that makes the food dangerous. This system is dynamic, and changes over time as we age and our health changes. Tying these all together are the many biochemical, physiological and metabolic processes with cells and tissues.

APPENDIX F

▼

PERMISSIONS:

Dear Anne,
You have permission to use the primer text and the photo of the DNA molecule for your book on DNA testing for genealogists. Please prominently credit
the U.S. Department of Energy Human Genome Program as the source for
both and also include our website for more information on the Human
Genome Project and its applications: www.ornl.gov/hgmis. Yes, you can use
the *Dictionary of Genetic Terms* at the end of the online version of the Primer.
Please provide source citations when you use each part of the document. We
would appreciate having a copy of your book when it is completed.
Sincerely,
Denise Casey

Denise K. Casey
Science Writer/Editor
Human Genome News
Human Genome Management Information System
Oak Ridge National Laboratory
1060 Commerce Park, MS 6480
Oak Ridge, TN 37830
865/574-0597; Fax: 865/574-9888; Email: caseydk@ornl.gov
HGMIS World Wide Web URL: http://www.ornl.gov/hgmis
Sponsor: U.S. Department of Energy

Harry Ostrer, M.D.
Professor of Pediatrics, Pathology, and Medicine
Director, Human Genetics Program
New York University School of Medicine
550 First Avenue, MSB 136

New York, NY 10016
tel 212 263-7596
fax 212 263-3477

————-Original Message————-
From: Frederic Abramson
To: Anne Hart
Sent: Wednesday, July 23, 2003 2:21 PM
Subject: Re: Permissions

Hi Anne

I would be honored to have the article included, and the quote about the
viruses.

Fredric

Index

0-595-29067-1

www.ingramcontent.com/pod-product-compliance
Lightning Source LLC
Chambersburg PA
CBHW020724180526
45163CB00001B/96